Modeling of Responsive
SUPPLY CHAIN

IIT Kharagpur Research Monograph Series

Published Titles:

Modeling of Responsive Supply Chain, *M.K. Tiwari, B. Mahanty, S. P. Sarmah, and M. Jenamani*

Micellar Enhanced Ultrafiltration: Fundamentals & Applications, *Sirshendu De and Sourav Mondal*

IIT KHARAGPUR RESEARCH MONOGRAPH SERIES

Modeling of Responsive
SUPPLY CHAIN

M.K. Tiwari • B. Mahanty
S. P. Sarmah • M. Jenamani

CRC Press
Taylor & Francis Group
Boca Raton London New York

CRC Press is an imprint of the
Taylor & Francis Group, an **informa** business

CRC Press
Taylor & Francis Group
6000 Broken Sound Parkway NW, Suite 300
Boca Raton, FL 33487-2742

First issued in paperback 2019

© 2013 by Taylor & Francis Group, LLC
CRC Press is an imprint of Taylor & Francis Group, an Informa business

No claim to original U.S. Government works

ISBN-13: 978-1-4665-1034-0 (hbk)
ISBN-13: 978-0-367-38096-0 (pbk)

Library of Congress Cataloging-in-Publication Data

Modeling of responsive supply chain / M.K. Tiwari ... [et al.].
    p. cm. -- (IIT Kharagpur research monograph series)
    Summary: "Addressing various aspects of supply chain management, this book describes the coordination between various elements in supply chain and optimizes the problem using both conventional and evolutionary approaches. It considers different models in the supply chain such as the transportation model, facility location model, assignment model, and planning and scheduling models. The text presents diverse technologies like RFID tags for detection of flow of particular item in the supply chain network. It also addresses the use of artificial intelligent optimization techniques in different types of supply chain problems and the use of specific coordination mechanisms and different analytical models"-- Provided by publisher.
    Includes bibliographical references and index.
    ISBN 978-1-4665-1034-0 (hardback)
    1. Business logistics. I. Tiwari, Manoj.

HD38.5.M585 2012
658.701'1--dc23   2012015725

Visit the Taylor & Francis Web site at
http://www.taylorandfrancis.com

and the CRC Press Web site at
http://www.crcpress.com

Contents

List of Figures

List of Tables

About the Series

IIT Kharagpur had been a forerunner in research publications and this monograph series is a natural culmination. Empowered with vast experience of over sixty years, the faculty now gets together with their glorious alumni to present bibles of information under the *IIT Kharagpur Research Monograph Series*.

Initiated during the Diamond Jubilee Year of the Institute, the series aims at collating research and developments in various branches of science and engineering in a coherent manner. The series, which will be an ongoing endeavour, is expected to be a source reference to fundamental research as well as to provide directions to young researchers. The presentations are in a format that can serve as stand-alone texts or reference books.

> The specific objective of this research monograph series is to encourage the eminent faculty and coveted alumni to spread and share knowledge and information to the global community for the betterment of mankind.

The Institute

Indian Institute of Technology Kharagpur is one of the pioneering Technological Institutes in India, and it is the first of its kind to be established immediately after the independence of India. It was founded in 18 August, 1951, at Hijli, Kharagpur, West Bengal, India. The IIT Kharagpur has the largest campus of all IITs, with an area of 2,100 acres. At present, it has 34 Departments and Centers and Schools and about 10,000 undergraduate, postgraduate, and research students with a faculty strength of nearly 600; the number of faculty is expected to double within approximately five years. The faculty and the alumni of IIT Kharagpur are having wide global exposures with the advances of Science and Engineering. The experience and the contributions of the faculty, students, and the alumni are expected to get exposed through this monograph series.

More on IIT Kharagpur is available at www.iitkgp.ac.in

Preface

The competitive business environment is changing significantly due to globalization, shorter product life cycles, minimum cost and lead time, and new technologies. A market-driven business must react quickly to changes in customer preferences, technology, and competition. It can reposition itself in its value chain or reengineer in such a manner so that flows of goods, information, and funds will facilitate its business processes. These options require the company to be responsive and provide long-term viability. However, it must also be capable of mobilizing the mass of a dinosaur.

One of the competition-winning factors in this field is the improvement of supply chain performance through responsiveness. Not only does responsiveness contain several criteria such as effectiveness, efficiency, quality, innovation, or flexibility, which need to be balanced within each firm, but also there should be a consistent alignment of these criteria in the entire supply chain. Therefore, a common responsive supply chain approach is required enabling companies to manage supply chain strategies through a fully integrated system of business improvement methodologies.

This monograph provides a valuable insight into novel concepts of responsive supply chain management. We intend to provide state-of-the-art concepts, models, and frameworks that would help in understanding how an adequate response to market forces can be shaped. The discussions are targeted to meet the requirement of academia as well as industry. A variety of illustrative examples have been provided to enhance the understanding of the readers. Also, new avenues of research have been identified. We have organized five chapters of this monograph covering the framework of supply chain responsiveness. The chapters include the use of optimization, coordination, system dynamics, and technology integration in achieving a higher responsiveness in a supply chain.

Chapter 1 discusses the overall importance of recognizing responsiveness as a metric of supply chain performance. In Chapter 2, we study domain interfaces in greater depth in terms of the importance of solving the optimization problem in supply chains to become more responsive. The importance of coordination through contracts in enhancing responsiveness of a supply chain is discussed in Chapter 3. Chapter 4 uses system dynamics methodology to achieve responsiveness while amalgamating management principles, control theory, and computer simulation. Finally, Chapter 5 presents the use of different types of technologies and also investigates its usefulness in building a better supply chain targeted to achieve higher responsiveness.

Manoj Kumar Tiwari, Biswajit Mahanty, S.P. Sarmah,
and Mamata Jenamani
IIT Kharagpur

Acknowledgments

The Indian Institute of Technology, Kharagpur, has been celebrating its Diamond Jubilee since August 2011 and, in this regard, there is a plan to publish a monograph on contemporary research areas. Supply chain management has assumed the crucial role of boosting industrial growth in the recent past due to proliferation of globalization in the manufacturing and transportation sectors. We have decided to address the most contemporary topics related to the supply chain that could have the potential to affect the free flow of goods and services among the nations and also to become a determinant in accessing the performance of logistic systems deployed to realize the above-mentioned objectives.

A responsive supply chain is one such emerging area that can significantly affect trade and business practices among the various organizations situated in different countries. This monograph provides a compact framework and illustrates how to make the supply chain more responsive. It is difficult to find such practical and conceptual approaches at one place with a detailed explanation. We have tried to address all such issues in this monograph on *Modeling of Responsive Supply Chain*. The present monograph not only acts as a treatise on the subject for scholars and graduate students but also as a compendium for practitioners while dealing with real-life problems.

We wish to express gratitude to the director professor D. Acharaya, IIT Kharagpur, for encouraging us to bring out this monograph to cover the diversity encountered in a responsive supply chain. The professor in charge of looking after the publication of the monograph, professor K.K. Ray deserves special mention from us for his variable assistance and support at different levels. We also thank Gagandeep Singh of CRC Press (Taylor & Francis Group) for advice and suggestions at different stages in preparing this monograph.

We would like to thank a number of our students and friends in the Department of Industrial Engineering and Management (IIT Kharagpur) who assisted us by providing technical support and checking of the manuscript and pointed out several flaws in an earlier version of the monograph. We offer our thanks to Akhilesh Kumar for his help in correcting the manuscript. We compliment our research scholars, namely, B.G. Menon, Priyabrata Mohapotra, Amit Kumar Sinha, Purushottam Lal Meena, M. Ramkumar, and others for their hard work in bringing out the monograph.

Authors

Biswajit Mahanty is currently professor and head of the Department of Industrial Engineering and Management at IIT Kharagpur. He has obtained his B.Tech (Hons) degree in Mechanical Engineering, and his M.Tech and Ph.D. degrees in Industrial Engineering and Management—all from IIT Kharagpur. To the present, Professor Mahanty has had a rich and varied professional career with over six years of industrial experience and about 22 years of teaching, research, and industrial consulting work experience. His areas of interest are system dynamics, operations research, information systems, and sup-

ply chain management. Professor Mahanty has guided eight doctoral and about 100 undergraduate and post-graduate level dissertations. He has also carried out 16 industrial consulting projects and five sponsored research projects. His publications have appeared in a number of well-known international journals. He has also taught in the School of Management at AIT, Bangkok for a brief period.

Manoj Kumar Tiwari works in the area of computational intelligence and its applications in manufacturing and operations management. His contribution has also been focused to solve production planning and control problems on manufacturing. Tiwari also addresses issues and concerns related to supply chain management. He has published more than 162 papers in respected international journals and has more than 80 papers in several national and international conferences. He is listed among top 20 most productive authors in the area of production and operations management as reported in the last 50 years

(published in a survey article in *International Journal of Production Economics*, 2009, 120, 540–551). He is associate editor of *International of System Science* (Taylor & Francis), *Journal of Intelligent Manufacturing* (Springer), *DSS, IEEE Transaction System Cybernetics and Man: Part-A*, and is area editor of *Computer and Industrial Engineering*. Dr. Tiwari is on the editorial board of several international journals such as *EJOR, IJPR, IJCIM, PPC, RCIM, IMechE, Part C, IJAMT,* and *JTS.* He is well known for grooming undergraduate and post-graduate students to carry out research and build up their careers through research. Several of his students are working as faculty members in different universities in the United Kingdom and United States and also in excellent global companies. Dr. Tiwari has a B.Tech (Mechanical Engg) from V.R.C.E, Nagpur, M.Tech (Production Engg) from MNIT, Allahabad, and Ph.D. from the Department of Production, Engg, Jadavpur University.

S.P. Sarmah obtained his Ph.D. degree from IIT Kharagpur, India and is currently working as an associate professor in the Department of Industrial Engineering and Management at IIT Kharagpur. Prior to teaching, he worked in industry for nearly four years. Dr. Sarmah's present research interests are in the areas of supply chain coordination, supply chain risk management, reverse logistics, production planning and control, inventory management, and project management. He has published papers in leading international journals such as *European Journal of Operational Research, International Journal of Production Economics, Mathematical and Computer Modeling: International Journal, Transportation Research Part E,* and *International Journal of Operational Research,* and his research articles are widely cited by researchers in the field. He is also a reviewer of many international journals. Apart from teaching and research, Dr. Sarmah is actively engaged with industry consultancy projects and has conducted short courses for academia and industry personnel.

Mamata Jenamani is currently working in the Department of Industrial Engineering and Management as an associate professor. Before that she worked in the Department of Computer Science and Engineering at the National Institute of Technology, Rourkela. She is a Ph.D. from IIT Kharagpurt and Post Doctoral Research from Purdue University. In 2005, she won an Emerald/EFMD Outstanding Doctoral Research Award in the category of Enterprise Applications of Internet Technology for her Ph.D. work. Her broad area of interest is information systems and e-business. Her current research interests include e-procurement, auctions, and application of information technology to supply chain systems.

1

Supply Chain Responsiveness

1.1 Introduction

In 1980s, researchers and practitioners started to look for new strategies to improve organizational competitiveness, which led to the inception of *supply chain management* (SCM). SCM began to attract interest in the mid-1990s and since then the essence of the discipline is widely acclaimed and accepted. Today, in every corridor of industry, academia, and governance, the concepts and applications of supply chains and their management implications have been widely practiced. A supply chain can be defined as "the delivery of enhanced customer and economic value through synchronized management of the flow of physical goods and associated information from source to consumption" (Jayaraman and Ross, 2003). Over the last three decades, SCM has spread its tentacles and grew to the status of an independent body of knowledge and a field of practice. A typical supply chain is composed of mainly three modules: inbound logistics, manufacturing operations, and outbound logistics. The first and last modules are respectively called "upstream" and "downstream" of the supply chain. An extended supply chain goes beyond these boundaries and tends to embrace the suppliers of suppliers on the upstream end and the entities of reverse logistics on the downstream end. At the beginning of the twenty-first century, companies have witnessed a period of change unparalleled in the history of the business world in terms of technological innovations, the globalization of markets, and more aggressive customer demands. The competitiveness in today's marketplace depends immensely on the ability of a company to handle the challenges of reducing manufacturing cycle time, reducing delivery lead time, increasing customer service level, and improving product quality. In order to stay competitive and for continual survival in the market, firms need to effectively handle fluctuations in an ever-changing market better than their competitors. This calls for building supply chains that are flexible and responsive enough to handle changing market and customer requirements.

Given prevailing competitiveness in the global market, firms are now fiercely exploring the potential of supply chains to increase their revenue, to differentiate themselves from competitors, to reduce their costs, and to add value to their supply chains and end customers. In particular, they are attempting to make supply chains more responsive to get their product to market faster while keeping the total supply chain cost at a minimum possible level. Catalan and Kotzab (2003) define responsiveness of a supply chain as the ability to respond and manage time effectively based on the ability to read and understand actual market signals. Christopher (2000), Fisher (1997), and Grossmann (2005) have commented that responsive supply chain is an essential strategy to gain competitive advantage. According to Hines (1998), supply chains that are more responsive can switch to a new generation of product within a shorter time period and are thus able to achieve an edge over their competitors in terms of capturing market shares.

In general most of the supply chains at least intend to be responsive. Now, the inevitable question arises: How challenging would it be to embrace responsiveness in supply chains while accounting for effective design and management, delivery of product variety, high quality, short lead times, high plant and volume flexibility, network structure, and inventory control? To do this effectively, supply chains may seek various alternatives such as, for example, (1) reducing production time via lot splitting or decreasing setup time; (2) investing in either finished products or raw materials safety stocks to buffer disruptions coming from the downstream or upstream, respectively; (3) opening retail outlets near the customer's market; (4) shipping via faster transportation modes; and (5) dealing with firms resorting to highly reliable suppliers. Furthermore, to gain critical competitive advantage, firms also need to understand their potential weak spots while assessing the implementation of new policies to be responsive (Stevenson, 2007).

1.2 Framework for Supply Chain Responsiveness

Figure 1.1 shows a framework for supply chain responsiveness. A responsive supply chain has the ability to respond to a wide range of quantitative demands, providing a high service level with a shorter lead time, handling a large variety of products, and handling uncertainty in supply. The framework is based on six major factors: (1) strategic planning, (2) virtual enterprises, (3) knowledge and information technology, (4) supply chain integration, (5) external determinants, and (6) operational factors. Integration of complementary core competencies and strategic alliance are necessary in developing virtual enterprises. A responsive supply chain can be achieved by forming good strategic alliances based on mergers and acquisitions with the objective of obtaining the required services. Apart from strategic alliance, corporate and business strategies, global outsourcing, type of market and products, location, government policies, and environmental regulations are other important factors. Tracey et al. (1994) suggested that any supply chain will be truly responsive if its parameters can move even more quickly with the efficient utilization of existing equipment, existing facilities, and existing designs. Review of enabling technologies suggested that the selection of technologies depends upon strategies that are selected to meet varying market requirements. ERP (enterprises resource planning) systems such

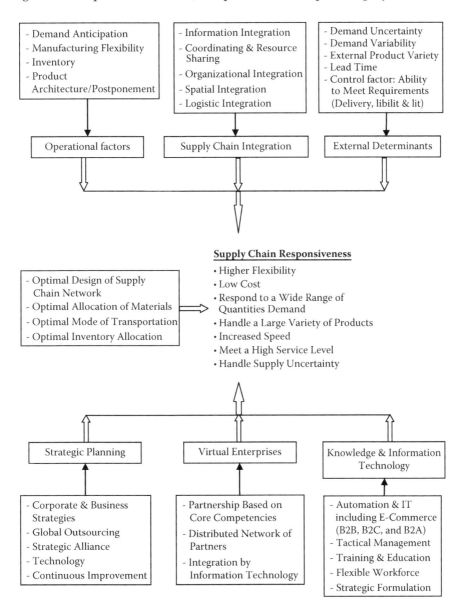

FIGURE 1.1
Framework for supply chain responsiveness.

as SAP, Oracle, Peoplesoft, BAAN, and JD Edwards can be used to achieve effective integration in a responsive supply chain.

The accurate anticipation of demand can help supply chains to respond to customer requirements faster. Manufacturing flexibility can directly reduce the production lead times and changeover times for products in the supply chain. Inventory is often used as a buffer against uncertainty and is linked to the decoupling point, which is a common criterion for classifying supply chain strategies, such as build-to-order supply chain. Product architecture/ postponement can be determined to a large extent where the decoupling points can be placed and which achieves responsiveness in the chain. It also determines manufacturing and internal product variety and complexity. Information integration can help reduce internal demands amplification and eliminate delays due to slow information flows. Coordination and resource sharing can reduce delay and unnecessary activities in supply chains. Organizational integration can increase the responsiveness of the supply chain by reducing demand variability and uncertainty.

This monograph aims to document how responsiveness can be achieved in a supply chain through optimization, coordination, system dynamics, and technology integration. In Chapter 2, we present the importance of optimization in supply chains and how to use it to make a supply chain more responsive. To facilitate better understanding, we explain formulations and solution methodology by using a few interesting problems. Chapter 3 emphasizes the importance of coordination through contracts in enhancing responsiveness of a supply chain. Chapter 4 is formulated with the application of systems dynamics to achieve responsiveness while combining management principles, control theory, and computer simulation. Finally, Chapter 5 analyzes the use of different types of technologies such as radio frequency identification (RFID), electronic data interchange (EDI), e-commerce, e-procurement, RosettaNet, extended relationship management (XRE), Partner Interface Process (PIP), business to business (B2B), and business to customer (B2C) in building a better supply chain network targeted to achieve higher responsiveness.

References

Catalan, M., and Kotzab, H. 2003. Assessing the responsiveness in the Danish mobile phone supply chain. *International Journal of Physical Distribution and Logistics Management*, 33(8), 668–685.

Christopher, M. 2000. The agile supply chain competing in volatile markets. *Industrial Marketing Management*, 29, 37–44.

Fisher, M.L. 1997. What is the right supply chain for your product? *Harvard Business Review*, 75(2), 105–116.

Grossmann, I.E. 2005. Enterprise-wide optimization: A new frontier in process systems engineering. *AIChE Journal*, 51, 1846–1857.

Hines, P. 1998. Benchmarking Toyota's supply chain: Japan vs. UK. *Long Range Planning*, 31(6), 911–918.

Jayaraman, V., and Ross, A. 2003. A simulated annealing methodology to distribution network design and management. *European Journal of Operational Research*, 144, 629–645.

Stevenson, W.J. 2007. *Operations Management*. 9th International ed. Boston, MA: McGraw-Hill Irwin.

Tracy, M.J., Murphy, J.N., Denner, R.W., Pince, B.W., Joseph, F.R., and Pilz, A.R. 1994. Achieving agile manufacturing, part II. *Automotive Engineering* (Warrendale, PA), 102(12), 13–17.

2

A Supply Chain Optimization Perspective to Achieve Responsiveness

2.1 Introduction

The essence of a supply chain lies in its beauty and capability of its partners working as a "coherent whole" to reap the benefits of cooperation and coordination, though it is aptly said that a supply chain can be as strong as its weakest link. This signifies the need and importance of building a strong, reliable, and robust supply chain. In a nutshell, the success or failure of any supply chain greatly depends on its structure and composition. Hence, restructuring of a supply chain is not a trivial issue; thus, the supply chain needs to be optimized, from time to time, to meet with market and customer requirements by embracing responsiveness. Therefore, the focus of this chapter is to introduce the importance of optimization in making a responsive supply chain.

Supply chain optimization is one broad area in SCM that deals with the application of processes and tools to ensure the optimal operation of various components and modules used in building a supply chain under diverse conditions. This includes ensuring the right inflow of the right materials on one end and the efficient and effective distribution of goods on the other end, with flawless production of goods at the in-between link. Thus, one needs to optimize constantly and continuously the costs, time, and quality involved in every aspect of inbound logistics, manufacturing operations, and outbound logistics. The supply chain optimization is typically dealt at three levels: strategic level, tactical level, and operational level. The strategic level deals with building an optimal structure of a supply chain network. The tactical and operational levels deal with middle-level and operational-level managerial decisions of a supply chain.

As previously mentioned, supply chain optimization is a vast discipline, and thus it is not possible to account for all the areas in one chapter. Thus, the scope of this chapter is to demonstrate the use of optimization in achieving a responsive supply chain by studying a few important supply chain problems. This chapter considers two important problems of any supply chain:

1. *Network design*: A crucial component of the planning activities of a manufacturing firm is the efficient design and operation of its supply chain. A supply chain is a network of suppliers, manufacturing plants, warehouses, and distribution channels organized to acquire raw materials, convert these raw materials to finished products, and distribute these products to customers. Strategic-level supply chain planning involves deciding the configuration of the network (i.e., the number, location, capacity, and technology of the facilities). The tactical-level planning of supply chain operations involves deciding the aggregate quantities and material flow for purchasing, processing, and distributing products. The strategic configuration of the supply chain is a key factor influencing efficient tactical operations, and therefore it has a long-lasting impact on the firm. Furthermore, the fact that the supply chain configuration involves the commitment of substantial capacity resources over long periods of time makes the supply chain network design problem extremely important (Santoso et al., 2005).

2. *Inventory management*: Leading manufacturing and distribution companies strive to make inventory management one of their core competencies. This fact ascribes pivotal importance to this field among the research community and industrial practitioners. The global nature of operations, volatile market conditions, and fierce competition make it imperative to devise new policies to maintain higher service levels at the lowest possible cost and time. Inventory management has thus become a vital part of any supply chain. Inventory management seeks to integrate the decision procedure for the whole supply chain and finds optimal inventory policies to meet the current aspiration levels. Efficient inventory management policies not only regulate the costs incurred but also make the firm more agile.

For both problems mentioned above, there are numerous traditional mathematical optimization models and solution methodologies available in the literature. But in this chapter, we have considered taking a detour from conventional solution methodology and described how various metaheuristics can be a good bet in finding optimal or near-optimal solutions. This is due to the fact that both these problems are combinatorial in nature and hence can be put into the category of non-deterministic polynomial-time (NP)-hard problems. The solution to these problems will have a large number of combinations and will be difficult and sometimes impossible to solve by the exact mathematical method in a given polynomial time. Therefore, we have used different metaheuristics techniques for solving such types of problems. The details of each of these metaheuristics are provided as they appear in the text.

In general, the objective of this chapter is to explain and record the genesis and development of various metaheuristics and their variants in solving different facets of supply chain optimization in view of increasing responsiveness.

2.2 Network Design Problem

By doing a comprehensive study of the supply chain design literature, we have identified a research gap in the development of the responsive supply chain models. Most of the studies deal with the production and distribution or logistics exclusively, but they are totally ignoring the benefits of the integrated approach (Huang et al., 2005; Romeijn et al., 2007). To achieve a responsive supply chain, our main emphasis should be to satisfy the customer demands at specified service levels incurring the lowest possible cost. With an integrated approach, a company may save logistics costs and production costs and simultaneously improve the service level by redesigning its supply chain network. Also, in order to be competitive and responsive in the current manufacturing scenario, companies must align with the suppliers and customers to streamline their operations and work together to achieve a level of responsiveness that contributes in its effective design (Yusuf et al., 2004). In light of the aforementioned views, this chapter deals with the design of a responsive supply chain that procures the benefits of an integrated approach at the strategic decision-making level.

The review highlighted some common disadvantages and serious limitations of the above-described approaches regarding the supply chain network design. First, the supply chain network designs are considered separately without comprehensive interlinking and dynamic network design consideration. Most of the discussed models consider the supply chain as static and deterministic systems, focus on a single or limited number of objectives, and cannot represent the ever-changing nature of the supply chain network design. There is a lack of feedback between supply chain operations and planning to ensure the supply's adaptation to a current execution environment. In this chapter we have suggested a model for designing the network of a new entrant supply chain under uncertain demand. The proposed model is formulated on the basis of multiechelon supply chain network design by considering the responsiveness as a design criterion for the supply chains by integrating customer sensitivity, processes, networks, and information systems. The five echelons of supply chain conceived in this formulation include suppliers, plants, distribution centers, cross-docks (CDs), and customer zones. The objective of this chapter is to integrate these five echelons with their conflicting objectives and identify the best suppliers and plants on the basis of selection criteria mentioned in Supply Chain Council reference model (SCOR, 2003). Thus, the multiechelon supply chain network becomes responsive not only in terms of volume and plant flexibility but also in terms of total supply chain cost.

2.2.1 Problem Formulation of Network Design

Integrated network design of production, distribution, and logistics activities in a responsive supply chain management is shown in Figure 2.1. Let us consider a design network that is conceived from a class of five echelons of supply chains including suppliers, plants, distribution centers, cross-docks, and customer zones. The concept of the network design problem has been adopted from Manish et al. (2008). This section highlights a multiechelon supply chain network design problem by considering the responsiveness as a design criterion for the supply chains by integrating customer sensitivity, processes, networks, and information systems. During the formulation of the model we have assumed three categories of Responsive Performance Index (RPI), namely, high (0.8 to 1), medium (0.6 to 0.8), and low (0.4 to 0.6). The objective is to integrate these five echelons with their conflicting objectives

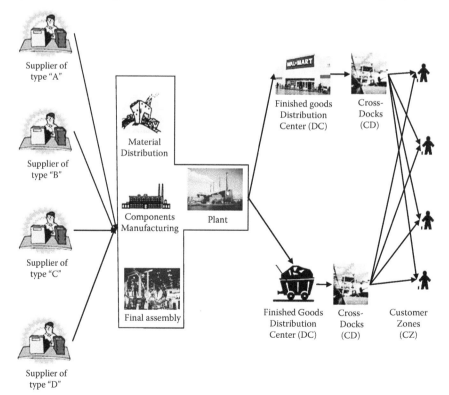

FIGURE 2.1 (See color insert.)
Integrated network of production, distribution, and logistics activities.

and identify the best suppliers and plants on the basis of selection criteria mentioned in the Supply Chain Council reference model. We also find the number of distribution centers and cross-docks to be opened for the demand of each customer zone.

$FDC_k \rightarrow$ Fixed cost for distribution center k

$HRG_l \rightarrow$ Lower bound of high responsive performance index (0.8)

$HRG_u \rightarrow$ Upper bound of high responsive performance index (1)

$I \rightarrow$ Total number of product to be manufactured; $i \in I$

$J \rightarrow$ Total number of cross-docks; $j \in J$

$K \rightarrow$ Total number of distribution centers; $k \in K$

$LRG_l \rightarrow$ Lower bound of low responsive performance index (0.4)

$LRG_u \rightarrow$ Upper bound of low responsive performance index (0.6)

$M \rightarrow$ Total number of customer zones; $m \in M$

$MRG_l \rightarrow$ Lower bound of medium responsive performance index (0.6)

$MRG_u \rightarrow$ Upper bound of medium responsive performance index (0.6)

$MNPV_{ip} \rightarrow$ Minimum production volume for product i at plant p

$MNTH_k \rightarrow$ Minimum throughput at distribution center k

$MXC \rightarrow$ Maximum number of cross-docks to be opened

$MXD \rightarrow$ Maximum number of distribution centers to be opened

$MXPV_{ip} \rightarrow$ Maximum production volume for product i at plant p

$MXTH_k \rightarrow$ Maximum throughput at distribution center k

$norm(.) \rightarrow$ Normalized objective (.)

$Obj \rightarrow$ Final integrated objective function

$P \rightarrow$ Total number of plants; $p \in P$

$PCP_p \rightarrow$ Production capacity for each plant p

$PCV_{rs} \rightarrow$ Production capacity of supplier s for raw material r

$UPCP_{ip} \rightarrow$ Unit production cost for product i at plant p

$UR_{ri} \rightarrow$ Utilization rate for each raw material r per unit of product i

$UTCD_{ikj} \rightarrow$ Unit transportation cost from distribution center k to cross-dock j for product i

$UTCP_{ipk} \rightarrow$ Unit transportation cost from plant p to distribution center k for product i

$UTCR_{rsp} \rightarrow$ Unit transportation cost from supplier s to plant p for raw material r

$\omega_{1m} \rightarrow$ Weight assigned by manufacturer to cost objective

$\omega_{2m} \rightarrow$ Weight assigned by manufacturer to distribution volume flexibility and plant flexibility

$RGCD_j \rightarrow$ Responsive level of cross-docks j

$RGP_p \rightarrow$ Responsive level of plant p

$RGS_s \rightarrow$ Responsive level of supplier s

$CCD_j \rightarrow$ Capacity of cross-dock j to handle product families

$CDC_k \rightarrow$ Capacity of distribution center k to handle product families

$CDF \rightarrow$ Cross-dock flexibility

$CSC_{mj} \rightarrow$ Cost to supply product i from cross-dock j which would be used by the customer zone m

$DCF \rightarrow$ Distribution center flexibility

$D_{mi} \rightarrow$ Demand from customer zone m for product i

$DVF_m \rightarrow$ Distribution value flexibility of customer zone m

$FCD_j \rightarrow$ Fixed operating cost to open cross-dock j

$FCP_p \rightarrow$ Fixed cost for plant p

$\lambda_{ip} \rightarrow$ Quantity of product i produced at plant p

$v_{ipk} \rightarrow$ Quantity of product i transported from plant p to distribution center k

$\phi_{rsp} \rightarrow$ Quantity of raw material r transported from supplier s to plant p

$PF \rightarrow$ Plant flexibility

$PVF_m \rightarrow$ Plant volume flexibility

$SUP_{ip} \rightarrow$ Standard unit at plant p per unit of product i

$R \rightarrow$ Total number of raw material to be supplied; $r \in R$

$S \rightarrow$ Total number of suppliers; $s \in S$

$SF \rightarrow$ Supplier flexibility

$SUDC_{ik} \rightarrow$ Standards units at distribution center k per unit of product i

$TC_m \rightarrow$ Total cost of m customer zone

$UCR_{rs} \rightarrow$ Unit cost of raw material r for supplier s

$UCT_{ik} \rightarrow$ Unit cost of throughput (handling and inventory) for product i at distribution center k

Decision Variables

$$X_p = \begin{cases} 1, & \text{if plant } p \text{ is open} \\ 0, & \text{otherwise} \end{cases}$$

$$Y_k = \begin{cases} 1, & \text{if distribution center } k \text{ is open} \\ 0, & \text{otherwise} \end{cases}$$

$$Z_j = \begin{cases} 1, & \text{if cross-dock } j \text{ is open} \\ 0, & \text{otherwise} \end{cases}$$

$$R_{jki} = \begin{cases} 1, & \text{if cross-dock } j \text{ is assigned to distribution} \\ & \text{center } k \text{ for product } i \\ 0, & \text{otherwise} \end{cases}$$

$$A_{mji} = \begin{cases} 1, & \text{if customer zone } m \text{ is assigned to cross-dock } j \\ & \text{for product } i \\ 0, & \text{otherwise} \end{cases}$$

$$N_s = \begin{cases} 1, & \text{if supplier } s \text{ is selected} \\ 0, & \text{otherwise} \end{cases}$$

The total supply chain cost will be composed of the following four different types of costs:

1. The raw material purchase cost and transportation cost from suppliers to plants

$$\sum_{p=1}^{P}\sum_{s=1}^{S}\sum_{r=1}^{R}\left(UTCR_{rsp}+UCR_{rs}\right)\varphi_{rmp} \tag{2.1}$$

2. The fixed and variable costs associated with plant operations

$$\sum_{p=1}^{P}FCP_p \times X_p + \sum_{i=1}^{I}\sum_{p=1}^{P}UPCP_{ip} \times \lambda_{ip} + \sum_{i=1}^{I}\sum_{p=1}^{P}\sum_{k=1}^{K}UPCP_{ipk} \times v_{ipk} \tag{2.2}$$

3. The variable cost of handling and inventory of products at distribution centers (DCs) and transportation of products from plant to DCs

$$\sum_{k=1}^{K}FDC_k \times Y_k + \sum_{i=1}^{I}\sum_{k=1}^{K}\sum_{j=1}^{J}UTCD_{ikj} \times R_{jki}$$

$$+\sum_{m=1}^{M}\sum_{i=1}^{I}\sum_{j=1}^{J}\sum_{k=1}^{K}UCT_{ik} \times D_{mi}R_{jki} \times A_{mji} \tag{2.3}$$

4. The transportation cost to ship product families from DCs to cross-docks and distribution costs to ship product family from cross-docks to customer zone

$$\sum_{j=1}^{J} FCD_j \times Z_j + \sum_{m=1}^{M}\sum_{i=1}^{I}\sum_{j=1}^{J} CSC_{mj} \times A_{mji} \tag{2.4}$$

Flexibility of the supplier's capacity

$$SF = \sum_{i=1}^{I}\sum_{r=1}^{R}\sum_{s=1}^{S}\left(PCV_{rs} \times N_s - D_{mi}\right) \tag{2.5}$$

Flexibility of the plant's capacity

$$PF = \sum_{i=1}^{I}\sum_{p=1}^{P}\left(PCP_p \times X_p - D_{mi}\right) \tag{2.6}$$

Flexibility of the distribution center's capacity

$$DCF = \sum_{i=1}^{I}\sum_{k=1}^{K}\left(CDC_k \times Y_k - D_{mi}\right) \tag{2.7}$$

Flexibility of the cross-dock's capacity

$$CDF = \sum_{i-1}^{I}\sum_{j=1}^{I}\left(CCD_j \times Z_j - D_{mi}\right) \tag{2.8}$$

Distribution volume flexibility

$$DVF_m = \min\left(SF, PF, DCF, CDF\right) \tag{2.9}$$

Plant volume flexibility

$$PVF_m = \sum_{p=1}^{P}\left(X_p \times PCP_p - \sum_{i=1}^{I}\lambda_{ip} \times SUP_{ip}\right) \tag{2.10}$$

The domains of cost and flexibility are different; therefore, a normalized approach is adopted to integrate both objective functions. The overall objective function consisting of normalized cost, distribution volume flexibility, and plant flexibility can be mathematically represented as

$$Mini\,Obj = \sum_{m=1}^{M} \left[\left(\omega_{1m} \times norm(TC_m) \right) - \omega_{2m} \times \{norm(DVF_m) + norm(PVF_m)\} \right] \quad (2.11)$$

where

$$\omega_{1m} + \omega_{2m} = 1 \quad (2.12)$$

The availability of required quantities of raw material with the supplier capability can be checked by Equation (2.13):

$$\sum_{p=1}^{P} \phi_{rsp} \leq PCV_{rs} \quad (2.13)$$

Equation (2.14) ensures that the raw material supplied by the supplier matches the production requirements:

$$\sum_{s=1}^{S} UR_{ri} \times \phi_{rsp} \leq \sum_{s=1}^{S} PCV_{rs} \quad (2.14)$$

$$\sum_{i=1}^{I} SUP_{ip} \times \lambda_{ip} \leq PCP_p \times X_p \quad (2.15)$$

Equation (2.15) shows that the total production quantity of products to be manufactured should not exceed the plant capacity.

The quantity of products to be produced at a plant should be within the minimum and maximum production capacities. This constraint is checked by Equation (2.16). Equation (2.17) enforces the minimum and maximum throughput capacities for DCs and ensures that customer zone assignments can be made only to open DCs.

$$MNPV_{ip} \times X_p \leq \lambda_{ip} \leq MXPV_{ip} \times X_p \quad (2.16)$$

$$MNTH_k \times Y_k \leq \sum_{i=1}^{I} \sum_{m=1}^{M} SUDC_{ik} \times D_{mi} \times R_{jki} \times A_{mji} \leq MXTH_k \times Y_k \quad (2.17)$$

$$\sum_{j=1}^{J} A_{mji} = 1 \quad (2.18)$$

Equation (2.18) shows that each customer zone must be assigned to exactly one cross-dock.

Equation (2.19) ensures that the amount transported from a plant is equal to the quantity of products available at that plant:

$$\lambda_{ip} = \sum_{k=1}^{K} v_{ipk} \tag{2.19}$$

Further, Equation (2.20) checks if all the demand requirements are satisfied:

$$\sum_{p=1}^{P} \sum_{k=1}^{K} v_{ipk} = \sum_{m=1}^{M} D_{mi} \tag{2.20}$$

Similarly, the demand requirements at each distribution center are checked by Equations (2.21) and (2.22):

$$\sum_{p=1}^{P} v_{ipk} = \sum_{m=1}^{M} D_{mi} \times R_{jki} \times A_{mji} \tag{2.21}$$

$$\lambda_{ip}, v_{ipk}, \phi_{rsp} \geq 0 \tag{2.22}$$

Equation (2.23) restricts cross-docks to be assigned to only open distribution centers:

$$R_{jki} = Z_j \tag{2.23}$$

Equation (2.24) ensures the distribution center's capacity restrictions:

$$\sum_{m=1}^{M} \sum_{i=1}^{I} d_{mi} \times A_{mji} \leq CCD_j \tag{2.24}$$

Equation (2.25) represents the capacity restriction for cross-docks:

$$\sum_{m=1}^{M} \sum_{i=1}^{I} R_{jki} \times CCD_j \leq CDC_k \tag{2.25}$$

Equation (2.26) ensures that customer demand for products is satisfied by open cross-docks:

$$A_{mji} \leq Z_j \tag{2.26}$$

Similarly, constraint given in Equation (2.27) checks that only open distribution centers will have product flow through its assigned cross-docks to customer zones:

$$R_{jki} = Y_k \tag{2.27}$$

Because capital availability is limited to every firm, it is essential to check whether the cost incurred in opening DCs and CDs should not exceed the available capital. This constraint is ensured by Equations (2.28) and (2.29):

$$\sum_{j=1}^{J} Z_j = MXC \tag{2.28}$$

$$\sum_{k=1}^{k} Y_k \leq MXD \tag{2.29}$$

As mentioned earlier, responsiveness is categorized in three classes: high, medium, and low. Based on this classification, Equation (2.30) ensures that the combined responsiveness of the selected suppliers, plants, DCs, and CDs should be less than the upper bound of the high level of the performance index and greater than its lower bound. Similarly for medium- and low-level responsiveness, the performance indexes are given in Equations (2.31) and (2.32), respectively.

$$HRG_l \leq RGS_s + RGP_p + RGDC_k + RGCD_j \leq HRG_u \tag{2.30}$$

$$MRG_l \leq RGS_s + RGP_p + RGDC_k + RGCD_j \leq MRG_u \tag{2.31}$$

$$LRG_l \leq RGS_s + RGP_p + RGDC_k + RGCD_j \leq LRG_u \tag{2.32}$$

2.2.2 Solution Methodology

Owing to the complexity of the integer and mixed-integer linear or nonlinear optimization models, it is difficult to solve them by using the exact method of optimizations. That makes the large-scale integer program extremely difficult to solve using the exact method of optimizations. Let us consider a few examples to show how the number of alternative solutions become large when we increase slightly the size of the problem. Consider an assignment problem of optimally matching 10 suppliers to 10 manufacturers (on a one-to-one basis). The number of feasible combinations of matching can be calculated as follows:

1. Theoretically, there are $n! = 3.628 * 10^6$ different combinations (where, $n = 10$).
2. If $n = 11$ (a 10% increase in the number of candidates and jobs), the number of alternative matching increases by 1,100% (i.e., to $39.9 * 10^6$)
3. If $n = 12$ (a 20% increase), the number of alternative matching increases by $13.2 * 10^2$% (i.e., to $48 * 10^7$).

Now, consider the well-known traveling salesman problem (TSP) where a salesman wants to find the least costly (or shortest) route when visiting clients in n different cities, visiting each city exactly once before returning home. Although this problem seems very simple, it becomes extremely difficult to solve as the number of cities would increase.

In general, for an n city TSP, there are $(n - 1)!$ possible routes that the salesman can take. Table 2.1 shows the value of $(n - 1)!$ for several n.

The total number of operations may vary with problem size, and even for similar-sized problems; the number can differ from one instance to another. One way of evaluating the complexity of an algorithm is to count, in the worst case, the total number of operations performed.

Let us consider the following:

O = Operations count

Ω = Lower bound on computational time

θ = range of upper and lower bound on an algorithm's performance

The complexity of the solutions can be represented in terms of required *computational efforts*. For example, if A is an $m * n$ matrix, then according to the Gaussian elimination we can say safely that the system $A * X = b$ has a total number of operations required that is $(n^3/3) + (n^2/2)$.

For a sufficiently large n, the growth of the second term is insignificant in comparison to the growth of the first term. Therefore, the first term dominates the second. Hence, the complexity of Gaussian elimination is $O(n^3)$ for the matrix A.

The referred time complexity function measures the upper bound of the rate of growth in solution time as the problem size increases. An algorithm

TABLE 2.1

Alternative Solution in Traveling Salesman Problem (TSP)

N	$(n - 1)!$
3	2
5	24
9	40,320
13	$47.9 * 10^6$
17	$20.9 * 10^{12}$
20	$12.1 * 10^{16}$

is recognized as "good" (or of acceptable performance) if its worst-case complexity is bounded by a polynomial function of the problem's parameters. Any such algorithm is said to be a polynomial-time algorithm. Examples of polynomial-time bounds are $O(n^2)$, $O(nm)$, and $O(\log n)$.

An algorithm is said to be an exponential-time algorithm if its worst-case computational time grows as a function that cannot be polynomial bounded by the input length. Some examples are $O(2^n)$, $O(n!)$, and $O(n^{\log n})$. A polynomial-time algorithm is always preferred as it is asymptotically superior to any exponential-time algorithm, even in extreme cases. For example, n^{4000} is smaller than $n^{0.1 \log n}$ if n is sufficiently large (i.e., $n \geq 2^{100,000}$).

The complexity of algorithms for solving optimization problems has been briefly discussed above; however, optimization problems may also be classified into varying sets of comparable complexity, called *complexity classes*. The complexity class P is the set of decision problems that can be solved using a deterministic approach in polynomial time. This class corresponds to a group of problems that can be effectively solved, even in the worst of cases, using an intuitive approach or guess. A problem is identified as NP (nondeterministic polynomial) if its solution (if one exists) can be guessed and verified in polynomial time; nondeterministic means that no particular rule is followed to make the guess.

The NP-complete problems are the toughest problems to be encountered of the NP type, in the sense that they are the ones most likely not to belong to the P class. This is because any problem of type NP can be transformed, in polynomial time, into an instance of a specific NP-complete problem.

For instance, the optimization problem version of the traveling salesman problem is NP-complete. Therefore, any instance of any problem in NP can be transformed mechanically into an instance of the TSP, in polynomial time. As a result, if the TSP turned out to be in P, then $P = NP$! The traveling salesman problem is one of many such NP-complete problems. If any NP-complete problem is in P, then it would follow that $P = NP$. Unfortunately, many important problems have been shown to be NP-complete, and no single expedient algorithmic solution method for any of them has been derived.

The supply chain network design problem is an NP-hard in nature (Ibaraki and Katoh, 1988). The supply chain network design optimization model formulated in Section 2.2.1 is also combinatorial in nature and comes under the banner of a NP-hard problem, because it contains a large number of variables and constraints. The solutions to such problems will have a large number of combinations that cannot be solved by exact mathematical models in polynomial time. When the exact method of optimization is used to solved this type of problem where decision variables are too high or constraint conditions are too complex, then the degree of complexity of the problem increases abruptly and required computational time will be very long, resulting in low efficiency, and sometimes solutions may be trapped in the range of partial optimum solution.

Therefore, such problems can only be solved by a slew of metaheuristics to find near-optimal solutions. The various metaheuristics put to task include nature-inspired, bio-inspired, physical phenomenon–inspired, and social phenomenon–inspired metaheuristics. The popular nature-inspired meta-heuristics include genetic algorithms (GAs), particle swarm optimization (PSO), and Artificial Honey Bee Algorithm. Artificial immune system (AIS) is an example of a bio-inspired metaheuristic. Simulated annealing (SA) is a physical phenomenon–inspired metaheuristic. And Tabu search (TS) is a social phenomenon–inspired metaheuristic.

The numerous variants of these metaheuristics have been developed by different researchers and used to address diverse nature and conditions of supply chain optimization problems. An equally good number of attempts and instances can be found in the present literature where a variety of hybrid versions of these metaheuristics are effectively used to solve such kinds of problems.

The objective of this monograph is to explore the concept of metaheuristic to solve a supply chain optimization problem. The different variants of meta-heuristics can be utilized to solve such types of problems. But for simplicity and brevity, we have used a few metaheuristic techniques to solve this problem. In the next section we discuss how simulated annealing will work to find optimal or near-optimal solutions of the above-mentioned optimization model.

2.2.3 Simulated Annealing Background

Simulated annealing (SA) was introduced as a heuristic approach to solve numerous combinatorial optimization problems to replace those schemes where the solution could get stuck on inferior solutions (Wilhelm and Ward, 1987). The relative success of the SA algorithm has been claimed by many researchers such as Kirkpatrick et al. (1992), Johnson et al. (1989), Golden and Skiscim (1986), Eglese (1990), and Yogeswaran et al. (2009). They have discussed the theoretical and practical aspects of SA in detail. The SA method is based on the simulation of thermal annealing of a typical heated solid. Ross and Jayaraman (2008) used SA methodology for a distribution network design problem.

SA is a random search approximation algorithm based on the simple concept of the physical annealing of solids. Metropolis et al. (1953) proposed the various steps of this algorithm which can be summarized in two parts: (1) starting with an infeasible solution and (2) repeatedly generating neighborhood solutions. A neighborhood solution is always accepted if it offers improvement in the objective function value, but the same is to be termed as worse if the solution has to be accepted with certain probability. The probability of acceptance of a neighborhood solution relies on the cooling temperature that is set in the beginning at a relatively higher value so that it allows the acceptance of a large proportion of the generated solution. The cooling temperature is slowly modified to minimize the probability of acceptance of inferior solutions. This prevents the algorithm from getting trapped in local optimum at an early stage. At each value of the cooling temperature, a number of moves are attempted, and the algorithm is stopped when a stopping criterion reaches a predetermined value.

Before the application of the SA algorithm, it is always desired that the following parameters of the algorithm be specified:

1. Internal configuration and solution space
2. Value of the control parameter in the beginning
3. Procedures for generating neighborhood solutions
4. Cooling rate
5. Stopping rule

2.2.3.1 Neighborhood Generation Procedure

Changing the sequence of operations will lead to change in the value of the objective function. To find the global optimal solution, the objective function passes through various stages of uphill and downhill moves. Therefore, to find a new operation sequence, a perturbation method has been used. In this case, the neighborhood generation method depends upon the schemes followed to perturb the initial solution.

Here, a perturbation scheme termed a *modified shifting scheme* (MSS), which is altogether different from the perturbation schemes proposed by researchers so far, is highlighted. Making small, local changes in the operation sequence may lead to generation of an infeasible neighborhood solution. According to MSS, the procedure of making local changes in the sequence may be such that it will generate the feasible sequence only. An advantage of this scheme is that it facilitates minimizing the search space and is helpful in reaching global optima value in a fewer number of iterations.

2.2.3.2 Cooling Schedule

Earlier researchers have adopted a cooling schedule with a constant cooling rate. Furthermore, it may not be necessary to spend much time at very high temperatures. The basic idea behind the method is to get downhill moves accepted, in order to search for better solutions (i.e., decrease the probability of getting trapped in local optima [local maxima]). As has been the case that the probability of acceptance of downhill moves is higher at higher temperature, a new method of varying the temperature, which was suggested by Sridhar et al. (1993), is used in this section. According to this method, the cooling rate is lower at the beginning, higher at the middle, and lower at the end, as is depicted in Figure 2.2.

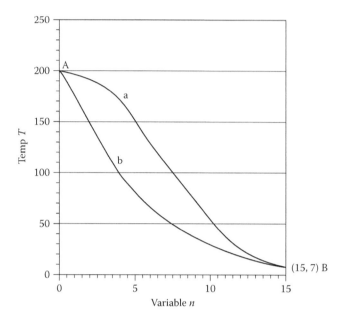

FIGURE 2.2
Temperature variation curves. (a) Proposed temperature variation curve. (b) Temperature variation curve for constant cooling rate proposed by Sridhar and Rajendran. (From Sridhar, J., and Rajendran, C. 1993. Scheduling in a cellular manufacturing system: A simulated annealing approach. *International Journal of Production Research*, 31, 2927–2945.)

For the curve (b) in Figure 2.2, the temperature variation governing the equation (constant cooling curve) is

$$T_b = (0.8)^n \times T_{initial} \qquad (2.33)$$

If $T_{initial} = 200°C$ and $T_{final} = 7°C$, then from Equation (2.33) we get $n = 15$ (approximately).

The proposed curve is cubic in nature and has maxima at (0, 200) and minima at (15, 7) in order to meet curve (b) at end points.

The equation used to govern the curve is as follows:

$$\text{At A,} \quad \frac{dT}{dn} = 0 \quad \text{and} \quad \frac{d^2T}{dn^2} < 0, \qquad (2.34)$$

and

$$\text{At B,} \quad \frac{dT}{dn} = 0 \quad \text{and} \quad \frac{d^2T}{dn^2} > 0 \qquad (2.35)$$

If T is the first differential, then

$$T = Kn(n - 15) \tag{2.36}$$

$$\Rightarrow dT = Kn(n - 15)dn \tag{2.37}$$

$$T = \int Kn(n - 15)\, dn \tag{2.38}$$

$$T = \frac{Kn^3}{3} - \frac{15Kn^2}{2} + C \tag{2.39}$$

Substituting the boundary condition in Equation (2.38) we get $K = 0.3431$ and $C = 200$.

Now the final governing curve will be

$$T = 0.1143 \times n^3 - 2.573 \times n^2 + 200 \tag{2.40}$$

2.2.3.3 Stopping Rule

The annealing schedule is usually stopped when it is observed that there is no change in configuration for a certain prescribed value of control parameter. Therefore, we accept a particular operation sequence that does not show an improvement in the local optima for certain consecutive values of the control parameter. Moreover, when the value of the temperature falls to a certain value (T_{final} in Figure 2.2 = 7°C), it has been observed that any further reduction in temperature would not be useful, as the probability of acceptance of an inferior solution is very low.

In the simulated annealing, the quality of the final solution is not affected by the initial guesses. The initial population for simulated annealing is generated randomly. Calculate the fitness value F_1 of the function. Generate the new population in the vicinity of the old population, and calculate the fitness value F_2. If the value of F_2 is smaller than the initial fitness F_1, then take the new population as the initial population. When F_2 is greater than F_1, then accepting or rejecting the point applies the Metropolis Criterion. According to the Metropolis Criterion for accepting the design point, there is a specified probability denoted by $\exp^{-(\Delta f/T)}$, where T is a parameter that is used as an initial temperature and $\Delta f = (F_2 - F_1)$.

2.2.4 Example of Network Design

This section deals with a five multiechelon supply chain network design problem that consists of suppliers, plants, distribution centers, cross-docks,

TABLE 2.2

Descriptions of Problem Design

Problem Instance	Suppliers (S)	Plants (P)	Distribution Centers (DC)	Cross-Docks (CD)	Customer Zones (CZ)	Number of Supply Chains to Be Designed (n)
1	20	20	15	15	20	3
2	15	15	15	15	15	3
3	12	12	12	12	12	3
4	12	12	10	10	10	3
5	10	10	10	10	10	3
6	8	8	8	8	8	3

and customers zones. The objective is to identify the optimal or near-optimal values of supply chain performance vecctors in terms of total supply chain cost and flexibility to achieve a responsive supply chain.

In this problem, we can define three classes of response, such as high, medium, and low, with performance indexes of, respectively, 0.8 to 1, 0.6 to 0.8, and 0.4 to 0.6. Also, we have to identify the optimal number of suppliers, plants, and distribution centers and provide the assignment of cross-docks to customer zones on the basis of problem descriptions as provided in Table 2.2. We have to design low, medium, and high responsive supply chains.

2.2.5 Implementations of Simulated Annealing

Pseudo code of SA is described as follows:

```
Generate initial population Xi
 Set the initial parameter (n = number of iteration,
  T as initial temperature (720),
  c as a temperature reduction factor (0.005)
   Repeat (until termination)
    While (iter<iter_max (50))
      {Calculate fitness value fi = f(Xi+1)
       Set iteration i = 1;
        cycle p = 1
         Generate new population Xi+1 in the vicinity of Xi
           If {fi - f(Xi+1) > 0}
            Accept new solution
             Else if {fi - f(Xi+1) ≤ 0}
              Select the solution with a probability exp-(Δf/T)
               Update iteration number i = i + 1
               Is number of iteration i = n
               Update the number of cycles as p = p + 1(until 5)
               Iter++(until 20)
```

```
            Reduce temperature
          End
       End
  End}
```

2.2.6 Results and Discussions

The computational tests presented in this section have been designed to evaluate the performance of the suggested SA algorithm. The problem instance conceived in this chapter corresponds to the dataset of a leading steel manufacturing company. It has been assumed that demand varies stochastically within each customer zone, and to satisfy the customer's need the steel company wishes to design n supply chain considering responsiveness as a key design criterion. A dataset related to the problem instance has been generated on a time horizon, and it considers the suppliers, plants, distribution centers, and cross-docks of varying responsive levels. Based on this classification, each echelon conceived here has a different responsive performance index (i.e., distribution in the range of 0.3 to 0.05).

As the responsiveness of the supply chain increases, it is desirable to have a large number of openings of cross-docks and distribution centers in any given time period. The problems considered here require different responsive supply chains to be designed after satisfying all of the given constraints. For a small size problem, say two to three suppliers, two to three plants, two to three DCs, and two to three CDs, an optimal or near-optimal solution can be found easily. However, complexity increases as the participating entities in each echelon of the supply chain approach a large size. In such scenarios, it would be difficult for any search technique to identify the optimal or near-optimal solutions for the given problem under consideration.

On account of the aforementioned scenario, a realistic problem design has been tested to validate the proposed model. Details of the problem design have been given in Table 2.1. It can be visualized from the dataset that problems of small and large scale are evaluated where the number of supply chains to be designed varies from one to three. The importance of incorporating cross-docks at the intermediate stage between DCs and CZs can be revealed by the fact that delivery of products from CDs to customer zones imparts responsiveness which is the prominent design criteria fixed by the manufacturers. Keeping in view the realistic scenario, we have assumed that for each problem instance, four products are required to be manufactured in different quantities, and each product requires 10 raw materials for its production. Specific weights are assigned to the normalized objectives of cost, plant flexibility, and volume flexibility in order to give freedom to the manufacturers in designing the supply chain of their own preference. Therefore, we introduce the cost-driven and flexibility-driven supply chains where weights associated to them are {0.8, 0.2} and {0.25, 0.75}, respectively.

The main motive of manufacturers is to design a supply chain of different levels of responsiveness and which should be either cost driven or flexibility driven. This implies that those supply chains that have high levels of responsiveness will certainly compromise with flexibility and cost as compared to those with low levels of responsiveness. However, the converse will be true for a low responsive supply chain as it would be a cost effective and flexibility enriched supply chain.

The suggested SA algorithm has been implemented over the problem design. In order to start the computational experiments, the first step is to fine-tune the algorithmic parameters so as to get the best possible results. The SA algorithm has been fine-tuned by varying the parameters range and thereafter the best set of operating values are found and detailed: number of iterations, 100; initial temperature, 720°C; temperature reduction factor, 0.005; and final temperature, 7°C. Table 2.2 shows the objective function value of cost, and flexibility for all problem instances which we have taken. Apart from these things, the optimal or near-optimal solution (i.e., suppliers, plants, distribution centers, and cross-docks for the problem instance in all three responsive supply chains, viz., low, medium, and high responsive supply chain network design) is shown in Table 2.3. The convergence of SA is shown in Figure 2.3.

TABLE 2.3

Low, Medium, and High Responsive Supply Chain Network Design

Problem Instance		Supplier	Plant	Distribution Center	Cross-Docks	Cost ($)	Flexibility
1	Low	11	12	9	7	683,407	3609
	Medium	5	4	3	6	1,001,213	3833
	High	4	4	3	2	1,401,692	3881
2	Low	6	7	6	9	631,292	3580
	Medium	6	5	4	4	943,428	3616
	High	3	3	5	2	1,222,403	3676
3	Low	7	6	6	8	506,026	3579
	Medium	2	3	4	2	756,987	3612
	High	3	3	2	2	954,851	3673
4	Low	7	7	5	6	356,431	3484
	Medium	3	2	3	2	558,973	3488
	High	2	3	3	2	762,143	3597
5	Low	4	5	4	5	307,916	3342
	Medium	4	2	2	2	509,703	3378
	High	4	3	6	3	714,063	3498
6	Low	4	3	3	3	241,324	3125
	Medium	3	3	2	2	364,162	3146
	High	1	4	2	3	519,872	3379

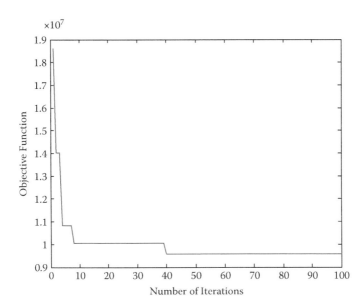

FIGURE 2.3
Convergence of simulated annealing.

2.2.7 Results Compare with Genetic Algorithm (GA)

We are going to solve the same problem using a genetic algorithm (GA), and last we will compare our previous results that we get from SA. In this section we are going to explain how the GA will be implemented in this problem.

2.2.7.1 Basic of GA

The usual form of genetic algorithm was described by Goldberg (1989). It is a metaheuristics based on the mechanism of natural selection and natural genetics. Let during iteration t, the GA contain a population (random solution) $P(t)$ of solutions $x_1^t, x_2^t, \ldots x_n^t$ (the population size n remains fixed). Each solution, x_n^t, is evaluated by computing $f\left(x_n^t\right)$, a measure of fitness of the solution. A new population, $P(t + 1)$, is then formed: we select a solution to reproduce on the basis of its relative fitness, and the selected solution is recombined using genetic operators (such as mutation and crossover) to form the new populations.

The mutation operators arbitrarily alter one or more components of a selected structure so as to increase the structural variability of the population. Each position of each solution in the population undergoes a random change with a probability defined by the mutation rate.

The crossover operator combined the features of two parent structures to form two similar offspring. It operates by swapping corresponding segments of a string representing the parent solution. For example, if the parents are represented by a five-dimensional vector,

A1	B1	C1	D1	E1

and

A2	B2	C2	D2	E2

then crossing the vector after the second component would produce the offspring

A1	B1	C2	D2	E2

and

A2	B2	C1	D1	E1

2.2.7.2 Pseudocode of GA

```
begin
  t = 0
    initialize P(t)
    evaluate P(t)
  while (nontermination-condition) do
  begin
    t = t + 1
    select P(t) from P(t - 1)
    recombined P(t)
    evaluate P(t)
  end
end
```

During implementation of the GA the parameters are simulated and set in the following manner: Population size, 50; Maximum number of generations, 200; crossover rate, 0.2; and mutation rate, 0.1. Using MATLAB®9 on an Intel Core II processor (1.2 MHz and 500 RAM), we obtained the results using both the algorithms for the given problem instance in six key scenarios. The comparative results for the scenarios based upon the performance of the two algorithms (SA and GA) are presented in Figure 2.4 through Figure 2.15 for all six problem instances.

Problem instance 1 (See Figures 2.4 and 2.5.)
Problem instance 2 (See Figures 2.6 and 2.7.)
Problem instance 3 (See Figures 2.8 and 2.9.)
Problem instance 4 (See Figures 2.10 and 2.11.)
Problem instance 5 (See Figures 2.12 and 2.13.)
Problem instance 6 (See Figures 2.14 and 2.15.)

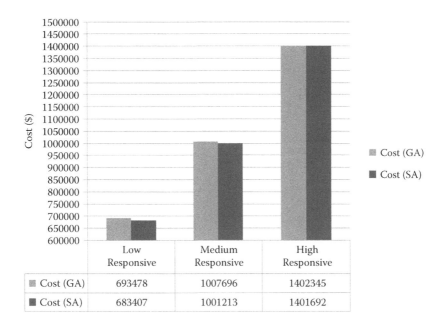

	Low Responsive	Medium Responsive	High Responsive
Cost (GA)	693478	1007696	1402345
Cost (SA)	683407	1001213	1401692

FIGURE 2.4 (See color insert.)
Comparison of results in terms of cost for problem instance 1.

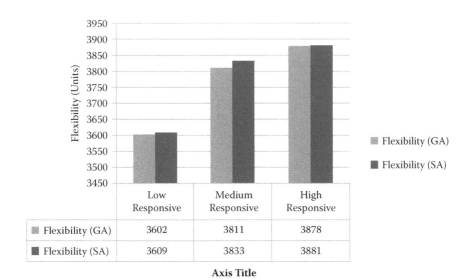

	Low Responsive	Medium Responsive	High Responsive
Flexibility (GA)	3602	3811	3878
Flexibility (SA)	3609	3833	3881

Axis Title

FIGURE 2.5 (See color insert.)
Comparison of results in terms of flexibility for problem instance 1.

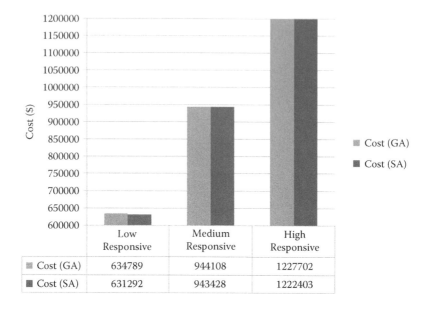

	Low Responsive	Medium Responsive	High Responsive
Cost (GA)	634789	944108	1227702
Cost (SA)	631292	943428	1222403

FIGURE 2.6 (See color insert.)
Comparison of results in terms of cost for problem instance 2.

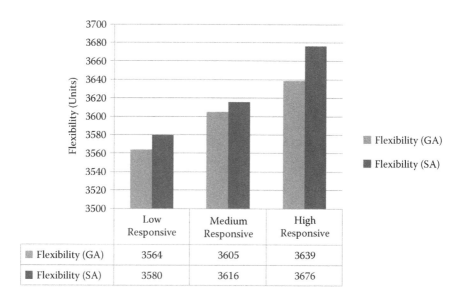

	Low Responsive	Medium Responsive	High Responsive
Flexibility (GA)	3564	3605	3639
Flexibility (SA)	3580	3616	3676

FIGURE 2.7 (See color insert.)
Comparison of results in terms of flexibility for problem instance 2.

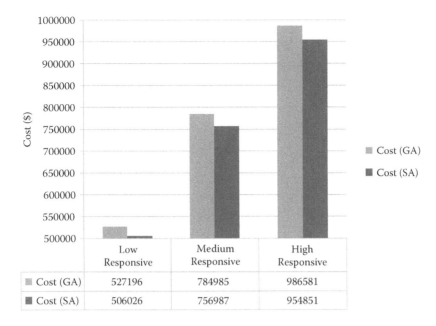

	Low Responsive	Medium Responsive	High Responsive
Cost (GA)	527196	784985	986581
Cost (SA)	506026	756987	954851

FIGURE 2.8 (See color insert.)
Comparison of results in terms of cost for problem instance 3.

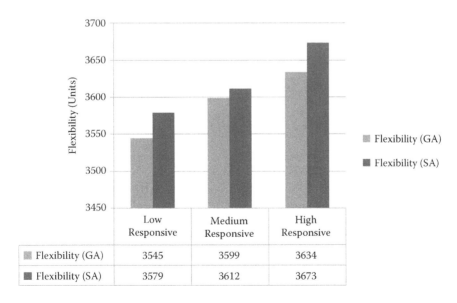

	Low Responsive	Medium Responsive	High Responsive
Flexibility (GA)	3545	3599	3634
Flexibility (SA)	3579	3612	3673

FIGURE 2.9 (See color insert.)
Comparison of results in terms of flexibility for problem instance 3.

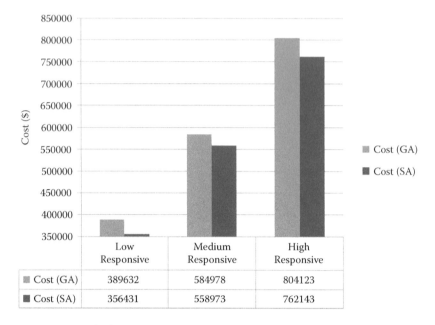

	Low Responsive	Medium Responsive	High Responsive
▮ Cost (GA)	389632	584978	804123
▮ Cost (SA)	356431	558973	762143

FIGURE 2.10 (See color insert.)
Comparison of results in terms of flexibility for problem instance 4.

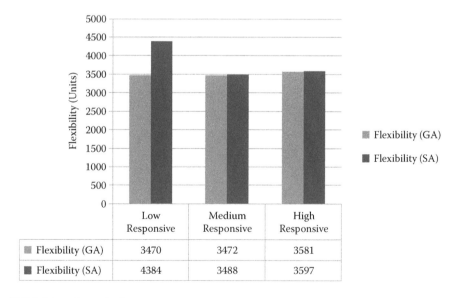

	Low Responsive	Medium Responsive	High Responsive
▮ Flexibility (GA)	3470	3472	3581
▮ Flexibility (SA)	4384	3488	3597

FIGURE 2.11 (See color insert.)
Comparison of results in terms of flexibility for problem instance 4.

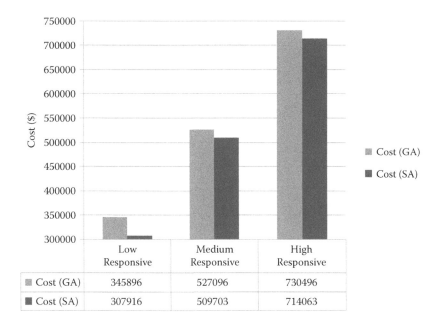

	Low Responsive	Medium Responsive	High Responsive
Cost (GA)	345896	527096	730496
Cost (SA)	307916	509703	714063

FIGURE 2.12 (See color insert.)
Comparison of results in terms of cost for problem instance 5.

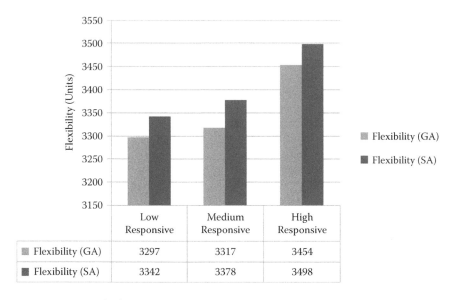

	Low Responsive	Medium Responsive	High Responsive
Flexibility (GA)	3297	3317	3454
Flexibility (SA)	3342	3378	3498

FIGURE 2.13 (See color insert.)
Comparison of results in terms of flexibility for problem instance 5.

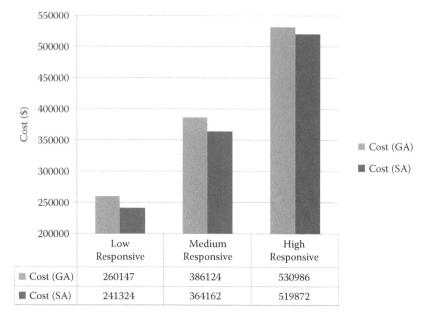

FIGURE 2.14 (See color insert.)
Comparison of results in terms of cost for problem instance 6.

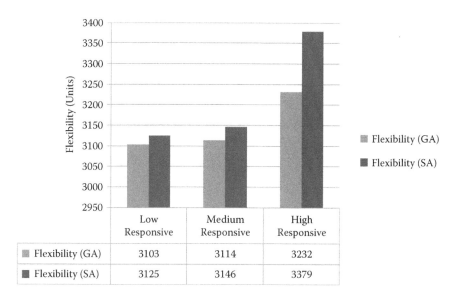

FIGURE 2.15 (See color insert.)
Comparison of results in terms of flexibility for problem instance 6.

This section clearly explains how to make a responsive supply chain in a five-echelon supply chain network by integrating the production, distribution, and logistics activities. A good network design can facilitate a higher value of flexibility. Supply chain network design can provide higher flexibility at higher total supply chain cost. The number of openings of cross-docks, distribution centers, and plants can be accommodated to achieve higher flexibility. That means that in supply chain network design, if we decrease the opening of cross-docks even then we can get higher flexibility by increasing the opening of either number of plants or number of distribution centers. It is quite clear from comparison of the results we get from SA and GA that SA suggests better results in terms of both total supply chain cost and flexibility in network design problem.

2.3 Responsiveness in Supply Chain Due to Inventory Management

In any industry a manufacturing and supply system is usually formed of a complex network of suppliers, fabrication/assembly locations, distribution centers, and customer locations, through which materials, components, products, and information flow (Ettl et al., 2000). Throughout this complex network there are some different sources of uncertainties that are associated with supplies not only in the form of its availability, quality, and delivery times, but also in demands such as arrival times, batch sizes, and its types. These uncertainties and other factors affect the performance of a system, including its service level in terms of delivery lead time, which in turn affects the bottom line of an enterprise in today's competitive environment. Among other things, inventories can be used to hedge uncertainties and achieve a specific service level. Because inventory placed at different locations usually incurs different costs and results in different service levels for the end customers, the efficient allocation and control of inventory assets facilitates enormous opportunities and, at the same time, poses a great challenge to many industries and service sectors.

An important issue in the management of supply chains and manufacturing systems is to control inventory costs at different locations throughout the system while satisfying the end-customer's service-level requirement. The global nature of operations, volatile market conditions, and fierce competition make it necessary to devise new inventory policies to maintain higher service levels in the lowest possible total operating cost and lead time. In today's global competitive environment, an efficient inventory planning is inevitable because it allows for more profit making and also provides higher system flexibility in terms of lead time for achieving a higher responsive supply chain. Its relative significance provides various objectives in a supply chain network with encapsulating different products and its lead time. Hence, we need an efficient inventory management to minimize the overall inventory in the system while meeting the required service level.

Inventory management becomes one of the most important core competency factors for any leading manufacturing and distribution company. Therefore, effective inventory management appears as a challenging task among the research community and industrial practitioners. Inventory management becomes an inseparable part of supply chain. Integration of the decision procedure for the entire supply chain and finding optimal inventory policies are major goals of inventory management. Effective and efficient management policy not only regulates the cost incurred but also makes the supply chain more responsive.

There are multiple performance metrics for assessment of any supply chain. Kleijnen and Smits (2003) critically analyzed these things by using simulation and then a balanced scorecard method to evaluate the overall performance of a firm. This problem can also be modeled as a multicriteria optimization problem. Several researchers have formulated mathematical models on inventory storage which encapsulate different useful managerial insights. The contributions of Chopra and Meindl (2001) demonstrated how the inventory at local retailers reduces the delivery times and facilitates a responsive supply chain.

2.3.1 Problem Formulation of Inventory Management

In this section, we discuss inventory planning for a three-echelon supply chain network consisting of manufacturer, warehouses, and retailers. Each facility is replenished by considering periodic review policies with optimal order up-to-quantities. Under this policy, when the review period for that particular facility comes and the inventory position is recorded to be below the reorder points, an order is issued to raise the stock up to a target level (S). Retailers are assigned to a particular warehouse, and this assignment is considered during network design phase. Keeping practical constraints under consideration, holding costs reduce as level of aggregation increases with each upstream layer of the network. The decision maker seeks to formulate the inventory policy for a fixed number of periods P under the forecasted demand or planned production data. The model optimizes inventory policy for the given set of periods and the forecasted data.

Notations:

$p \rightarrow$ Period

$i \rightarrow$ Warehouses

$j \rightarrow$ Retailers

$P \rightarrow$ Total period being considered

$I_{Wi} \rightarrow$ Inventory of warehouse

$I_{Rj} \rightarrow$ Inventory of retailers

$HC_{Rj} \rightarrow$ Holding cost for retailer j for one period

$HC_{Wi} \rightarrow$ Holding cost for warehouse i for one period

$OPC_W \rightarrow$ Order processing cost per order for warehouse

$OPC_M \rightarrow$ Order processing cost per order for manufactural

$OC_R \rightarrow$ Ordering cost per order for retailer

$OC_W \rightarrow$ Ordering cost per order for warehouse

$FTC \rightarrow$ Fixed transportation cost per order

$VTC(M,W_i) \rightarrow$ Variable transportation cost per unit transported from manufacturer to warehouse i

$VTC(W_i,R_j) \rightarrow$ Variable transportation cost per unit transported from warehouse i to retailer j

$RP_{Rj} \rightarrow$ Review period for retailer j

$RP_{Wi} \rightarrow$ Review period for warehouse i

$LT(M,W_i) \rightarrow$ Lead time from manufactural to warehouse j

$LT(W_j,R_i) \rightarrow$ Lead time from warehouse j to retailer i

$s_{Wi} \rightarrow$ Reorder level of inventory for warehouse i

$S_{Wi} \rightarrow$ Optimal order up-to-quantities for warehouse i

$s_{Rj} \rightarrow$ Reorder level of inventory for retailer j

$S_{Rj} \rightarrow$ Optimal order up-to-quantities for retailer j

$Q_R \rightarrow$ Total inventory order quantity of retailers

$Q_W \rightarrow$ Total inventory order quantity of warehouses

Maximum capacity constraints

$$S_{Rj} \leq S_{R\,max}, \quad j = 1,2,\dots,\text{ Number of retailers} \tag{2.41}$$

$$S_{Wi} \leq S_{W\,max}, \quad i = 1,2,\dots,\text{ Number of warehouses} \tag{2.42}$$

Review period

$$1 \leq RP_{Rj} \leq P, \quad j = 1,2,\dots,\text{Number of retailers} \tag{2.43}$$

$$1 \leq RP_{Wi} \leq P, \quad i = 1,2,\dots,\text{Number of retailers} \tag{2.44}$$

Minimum order quantity

$$S_{Rj} - s_{Rj} \geq Q_R, \qquad j = 1, 2, \dots, \text{Number of retailers} \qquad (2.45)$$

$$S_{Wi} - s_{Wi} \geq Q_W, \qquad i = 1, 2, \dots, \text{Number of warehouses} \qquad (2.46)$$

The maximum capacity constraints impose an upper limit to each facility's capacity to store merchandise and goods. Review periods have been subjected to practical constraints, as the review period has to be greater than or equal to one and the decision maker has selected a suitably long time span such that for normal functioning (every facility must review at least once) the review periods have to be less than or equal to number of periods, P. The minimum order quantity constraints impose a practical minimum order quantity as it may a case of inconvenience for the replenishing facility to entertain small orders. These constraints have now defined a bounded solution space for the optimization variables that can now be solved using a proposed algorithm. Here locations of warehouses are situated in geographically distant areas but locations of retailers are close to warehouses. Due to this location diversification, retailers would be supplied by a fixed warehouse. For keeping in mind the complexity of the problem, we are going to consider the manufacturing plant's production lead time and stock-out cost are equal to zero. Now we can easily say that no extra cost is incurred when backorders are met.

2.3.1.1 Minimizing Total Operating Cost

Total cost incurred by the supply chain includes ordering, processing, holding, and transportation costs by all participating facilities in the supply chain.

$$TC_{(Total\ cost)} = TC_{(Retailers)} + TC_{(Warehouses)} + TC_{(Transportation)} \qquad (2.47)$$

$$TC_{(Retailers)} = Ordering\ Cost + Holding\ Cost \qquad (2.48)$$

$$TC_{(Retailers)} = (no.\ of\ orders) \times OC_R + \sum_{j=1}^{no.\ of\ retailers} \sum_{p=1}^{P} HC_{Rj} \times I_{Rj} \qquad (2.49)$$

$$TC_{(Warehouses)} = Ordering\ Cost + Order\ Process\ Cost + Holding\ Cost \qquad (2.50)$$

$$TC_{(Warehouses)} = (no. of orders) \times OC_W + (no. of orders) \times OPC_W \qquad (2.50)$$

$$+ \sum_{i=1}^{no.\ of\ warehouses} \sum_{p=1}^{P} HC_{Wj} \times I_{Wi}$$

$$TC_{(Transportation)} = (no. of orders) \times FTC \qquad (2.51)$$

$$+ \sum_{i=1}^{no.\ of\ warehouses} (order\ quantity) \times VTC(M, W_i)$$

$$+ \sum_{i=0}^{no.\ of\ retailers} (order\ quantity) \times VTC(W_i, R_j)$$

2.3.1.2 Minimizing Weighted Lead Time

Lead time has been considered not in the form of transportation lead time but in the form of time between registration of demand and fulfillment of demand. Weighted average of the lead time, weighted by the total demand has been considered.

$$WLT_{(weighted\ lead\ time)} = \frac{\displaystyle\sum_{j=1}^{no.\ of\ retailers} \frac{\displaystyle\sum_{i=1}^{p} dp \times lp}{\displaystyle\sum_{p=1}^{p} dp}}{no. of\ retailers} \qquad (2.52)$$

where dp is demand in period p and l_p is lead time in period p.

2.3.2 Solution Methodology—Tabu Search

2.3.2.1 Tabu Search

Tabu search is a local search-based metaheuristic method that has been successfully applied to a wide class of NP hard optimization problems. Appropriate subject areas include bioengineering, finance, manufacturing, scheduling, and political districting. It was first presented by Glover (1986) and also sketched by Hansen (1986).

Tabu search uses a short-term memory structure called a *tabu-list*. A potential solution is marked as *tabu* so that the algorithm does not visit that possibility repeatedly. Tabu search starts with an initial solution. Algorithms based on Tabu search perform a neighborhood search (i.e., a local search) starting from a current solution to its best neighbor (the one with the best objective value among all examined candidates). Tabu search modifies the neighborhood structure of each solution as the search progresses. All the neighbors of a current solution are examined, and the best nonforbidden move is selected. Note that this move may decrease the quality of the solution but is necessary in order to increase the likelihood of escaping from local optimum "traps." A tabu list stores all the previously exploited moves or solutions that are now forbidden. The search continues until some stopping criterion has been satisfied.

To avoid cycling during the search process, the reverses of the last certain number of moves, formed as a tabu list, are prohibited or announced as tabu restricted for a certain number of iterations (i.e., the tabu duration). To prevent too rigorous parameter settings of the tabu restriction, some aspiration criteria are usually introduced which allow overriding the tabu restriction and thereby guide the search toward a promising region. Intensification and diversification strategies with Tabu search are also applied to emphasize and broaden the search in the solution space, respectively. Brief discussions about Tabu search can be found in the works of Chao (2002), Vilcot and Billaut (2008). Figure 2.16 shows the flowchart of Tabu search.

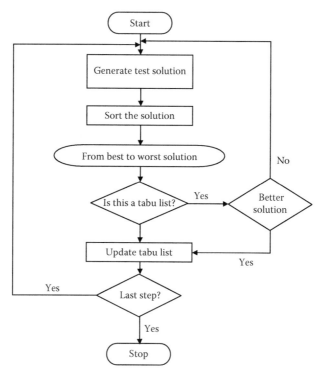

FIGURE 2.16
Flowchart of Tabu search.

TABLE 2.4

Parameters for Each Facility in the Supply Chain Network

Facility	Warehouse 1	Warehouse 2	Retailer 1	Retailer 2	Retailer 3	Retailer 4
S_{min} (units)	200	250	15	20	20	25
Maximum capacity (units)	1000	1200	100	120	140	150
Holding cost ($/unit/period)	4	4	40	40	30	30

The basic components of the Tabu algorithm are as follows:

1. *Configuration*: Coding of a solution
2. *Move*: Selected feasible direction of the search
3. *Set of candidate moves*: Feasible directions of the search
4. *Tabu restrictions*: The length of the tabu list
5. *Aspiration criteria*: Overriding the tabu restriction
6. *Stopping condition*: Terminating the search

2.3.3 Example of Inventory Management

Let there be two warehouses and four retailers whose minimum target level, maximum capacity, and holding costs are mentioned in Table 2.4.

Let us consider

Ordering cost = $1/order

Order processing cost = $2/order

Fixed transportation cost = $10/unit/order

The demand forecasts for determination of inventory positions during 20 span periods are shown in Figure 2.17 and Table 2.5.

In this problem we have assumed transportation cost as a variable quantity, and it is proportional to the lead time.

2.3.4 Implementation of Tabu Search

The pseudocode of Tabu search is described below.

```
cbest = c(S)
  do
  i = 1,maximum_iteration
  Δcbest=∞
  do
```

```
 j = 1,P
  do
   k = 1, ncolumn - P
    swap jth entry of IN with kth entry of out ⇒Yielding S̄ᵢ
    Calculate Δc = c(S̄ᵢ) - c(Si)
    if (Δc < Δcbest) then
     check tabu status of swap
    if(swaptabu) then
     check aspiration criteria
     if(c(S̄ᵢ) < cbest) then
      make swap best so far
      Si = S̄ᵢ
      Δbest = Δc
      else
       swap not allowed
      endif
     else
     make swap best so far
     Si = S̄ᵢ
      Δbest = Δc
     endif
    endif
   enddo
  enddo
  update tabu list with best swap
 if(c(Si) < cbest) then
  cbest = c(Si)
 endif
enddo
```

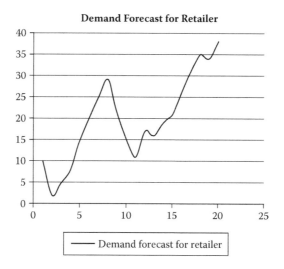

Demand Forecast for Retailer

Demand forecast for retailer

FIGURE 2.17
Demand forecasts for retailers.

TABLE 2.5

Demand Forecasts for Retailers

Time span (units)	1	2	3	4	5	6	7	8	9	10	11	12	13	14	15	16	17	18	19	20
Demand forecast (units)	10	2	5	8	15	20	25	29	21	15	11	17	16	19	21	26	31	35	34	38

2.3.5 Results and Discussion

We will reveal the results obtained by various weight factors associated with the weighted lead time. The results obtained from the Tabu search are described as follows:

> *Result 1* (when weighted factor of lead time is only 20%):
>
> Total cost = $33,889
>
> Weighted lead time = 26 units
>
> And the breakup of weighted lead time is shown in Table 2.6.
>
> In this situation, the assigned capacities of warehouses and retailers are provided in Table 2.7.

> *Result 2* (when weighted factor of lead time is only 30%):
>
> Total cost = $34,597
>
> Weighted lead time = 24 units
>
> And the breakup of weighted lead time is shown in Table 2.8.
>
> In this situation, the assigned capacities of warehouses and retailers are provided in Table 2.9.

TABLE 2.6

Distribution of Weighted Lead Time among Manufacturer, Warehouses, and Retailers

Facilities	Warehouse 1	Warehouse 2	Retailer 1	Retailer 2	Retailer 3	Retailer 4
Manufacturer	4	5				
Warehouse 1			4	5		
Warehouse 2					5	3

TABLE 2.7

Assigned Capacities of Warehouses and Retailers

Facilities	Warehouse 1	Warehouse 2	Retailer 1	Retailer 2	Retailer 3	Retailer 4
Capacity	263	353	17	22	24	50

TABLE 2.8

Distribution of Weighted Lead Time among Manufacturer, Warehouses, and Retailers

Facilities	Warehouse 1	Warehouse 2	Retailer 1	Retailer 2	Retailer 3	Retailer 4
Manufacturer	3	4				
Warehouse 1			4	5		
Warehouse 2					5	3

TABLE 2.9

Assigned Capacities of Warehouses and Retailers

Facilities	Warehouse 1	Warehouse 2	Retailer 1	Retailer 2	Retailer 3	Retailer 4
Capacity	408	289	21	28	25	49

Result 3 (when weighted factor of lead time is only 40%):

Total cost = $38,512

Weighted lead time = 18 units

And the breakup of weighted lead time is shown in Table 2.10.

In this situation, the assigned capacities of warehouses and retailers are provided in Table 2.11.

Result 4 (when weighted factor of lead time is only 50%):

Total cost = $38,667

Weighted lead time = 15 units

And the breakup of weighted lead time is shown in Table 2.12.

In this situation, the assigned capacities of warehouses and retailers are provided in Table 2.13.

TABLE 2.10

Distribution of Weighted Lead Time among Manufacturer, Warehouses, and Retailers

Facilities	Warehouse 1	Warehouse 2	Retailer 1	Retailer 2	Retailer 3	Retailer 4
Manufacturer	3	1				
Warehouse 1			3	4		
Warehouse 2					3	4

TABLE 2.11

Assigned Capacities of Warehouses and Retailers

Facilities	Warehouse 1	Warehouse 2	Retailer 1	Retailer 2	Retailer 3	Retailer 4
Capacity	246	602	15	39	27	60

TABLE 2.12

Distribution of Weighted Lead Time among Manufacturer, Warehouses, and Retailers

Facilities	Warehouse 1	Warehouse 2	Retailer 1	Retailer 2	Retailer 3	Retailer 4
Manufacturer	1	4				
Warehouse 1			2	4		
Warehouse 2					2	2

TABLE 2.13

Assigned Capacities of Warehouses and Retailers

Facilities	Warehouse 1	Warehouse 2	Retailer 1	Retailer 2	Retailer 3	Retailer 4
Capacity	253	420	65	24	23	64

Result 5 (when weighted factor of lead time is only 60%):

Total cost = $39,434

Weighted lead time = 11 units

And breakup of weighted lead time is shown in Table 2.14.

In this situation, the assigned capacities of warehouses and retailers are provided in Table 2.15.

Result 6 (when weighted factor of lead time is only 70%):

Total cost = $41,498

Weighted lead time = 10 units

And the breakup of weighted lead time is shown in Table 2.16.

In this situation, the assigned capacities of warehouses and retailers are provided in Table 2.17.

TABLE 2.14

Distribution of Weighted Lead Time among Manufacturer, Warehouses, and Retailers

Facilities	Warehouse 1	Warehouse 2	Retailer 1	Retailer 2	Retailer 3	Retailer 4
Manufacturer	1	4				
Warehouse 1			1	1		
Warehouse 2					1	3

TABLE 2.15

Assigned Capacities of Warehouses and Retailers

Facilities	Warehouse 1	Warehouse 2	Retailer 1	Retailer 2	Retailer 3	Retailer 4
Capacity	305	264	44	31	28	34

TABLE 2.16

Distribution of Weighted Lead Time among Manufacturer, Warehouses, and Retailers

Facilities	Warehouse 1	Warehouse 2	Retailer 1	Retailer 2	Retailer 3	Retailer 4
Manufacturer	2	2				
Warehouse 1			1	2		
Warehouse 2					2	1

TABLE 2.17

Assigned Capacities of Warehouses and Retailers

Facilities	Warehouse 1	Warehouse 2	Retailer 1	Retailer 2	Retailer 3	Retailer 4
Capacity	233	313	25	43	58	34

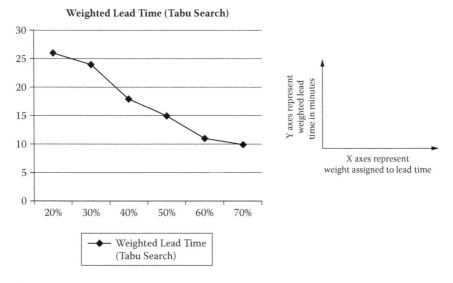

FIGURE 2.18
Weighted lead time according to Tabu search algorithm at various weights.

Figure 2.18 shows the weighted lead time according to the Tabu search algorithm at various assigned weights; Figure 2.19 shows the total cost at various weights.

2.3.6 Results by Implementing Particle Swarm Optimization (PSO)

We are going to solve the same inventory management problem by using PSO. In this section we are going to explain how the PSO has been implemented in this problem.

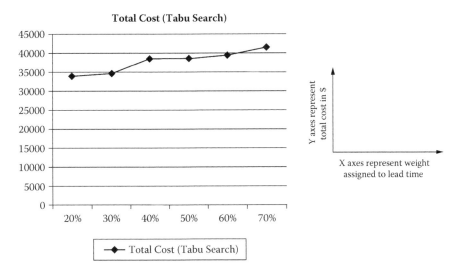

FIGURE 2.19
Total cost according to Tabu search algorithm at various weights.

Let us consider the following notations apart from previous notations:

$f(X) \rightarrow$ Objective function/fitness function with constraints

$X^{(l)} \rightarrow$ Lower bound

$X^{(u)} \rightarrow$ Upper bound

$g_j(X) \leq 0; j = 1, 2, 3, \ldots m$

$g_j(X) \rightarrow$ Constraints

$F(X) \rightarrow$ Objective function/fitness function without constraints

$X_j^{(i)} \rightarrow$ Position of the jth particle in iteration i

$V_j^{(i)} \rightarrow$ Velocity of jth particle in iteration i

$q_j(X) \rightarrow$ Magnitude of violation of the jth constraint

$\alpha, c, a, b \rightarrow$ Constants

$\varphi\left[q_j(X)\right] \rightarrow$ Continuous assignment function

$\gamma\left[q_j(X)\right] \rightarrow$ Power of the violated function

$$\gamma\left[q_j(X)\right] = \begin{cases} 1 & \text{if} \quad q_j(X) \leq 1 \\ 2 & \text{if} \quad q_j(X) > 1 \end{cases}$$

$c1 \rightarrow$ Cognitive (individual) learning rate

$c2 \rightarrow$ Social (group) learning rate

$r1 \& r2 \rightarrow$ Uniformly distributed random numbers in the range 0 & 1

$P_{best,j} \rightarrow$ Highest value of objective function $F\big[(X_j(i))\big]$,encountered by particle j

$Gest \rightarrow$ Highest value of the objecctive function $F\big[(X_j(i))\big]$

STEP 1: Let us consider size of the swarm (number of particles) = N

STEP 2: Generate "N" number of random initial population in between $X^{(l)}$ & $X^{(u)}$

Let us assume $X_1, X_1, \ldots X$

STEP 3: Calculate $q_j(X) = \max\big(0, g_j(X)\big);\quad j = 1, 2 \ldots, m$

STEP 4: Calculate $\varphi\big[q_j(X)\big] = a\left(1 - \dfrac{1}{e^{q_j(X)}}\right) + b$

We have taken $a = 150$; and $b = 10$

STEP 5: Calculate $H(X) = \displaystyle\sum_{j=1}^{m}\left\{\varphi\big[g_j(X)\big]\big[q_j(X)\big]^{\gamma[q_i(X)]}\right\}$

STEP 6: $C(i) = (ci)^{\alpha}$

We have taken $\alpha = 2$ and $c = 0.5$

STEP 7: Evaluate objective function $F\big[X_1(0)\big],\ F\big[X_2(0)\big],\ldots F\big[X_N(0)\big]$ by using $F(X) = f(X) + C(i) + H(X)$

STEP 8: Consider initial velocity; $V_1(0) = V_2(0) = \ldots = V_N(0) = 0$

STEP 9: Put $i = i + 1$ (loop start)

STEP 10: In the ith iteration evaluate

(A): $P_{best,j}$

(B): $P_{best,j}$

(C): $V_j(i) = V_j(i-1) + c_1 r_1\big[P_{best,j} - X_j(i-1)\big] + c_2 r_2\big[G_{best} - X_j(i-1)\big];$

$$j = 1, 2 \ldots, N$$

(D): $X_j(i) = X_j(i-1) + V_j(i);\quad j = 1, 2, \ldots, N$

(E): Evaluate $F\big[X_1(i)\big],\ F\big[X_2(i)\big],\ldots, F\big[X_N(i)\big]$

STEP 11: Check convergence criteria

```
        if convergence criteria satisfy
          break
        else
```

Go to STEP 9.

Figure 2.20 shows the weighted lead time according to PSO algorithm at various assigned weights. Figure 2.21 shows the total cost at various weights.

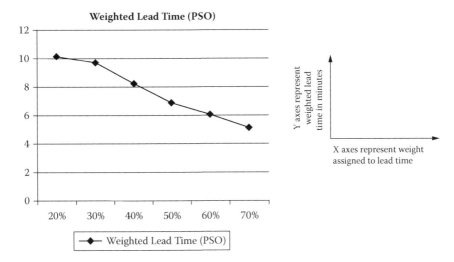

FIGURE 2.20
Weighted lead time according to particle swarm optimization algorithm at various weights.

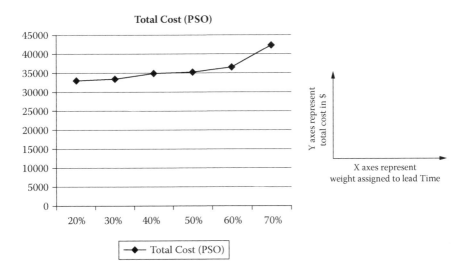

FIGURE 2.21
Total cost according to particle swarm optimization algorithm at various weights.

2.3.7 Comparison of Results between Tabu Search and PSO Algorithm

Figure 2.22 shows the comparison of weighted lead time according to PSO and Tabu search algorithms at various assigned weights. Figure 2.23 shows the comparison of total supply chain cost according to PSO and Tabu search algorithms at various assigned weights.

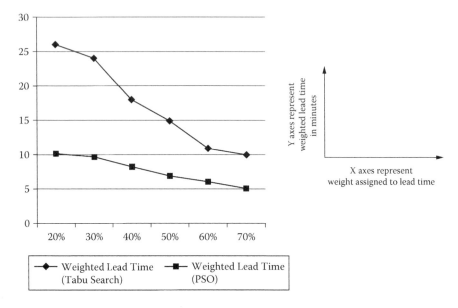

FIGURE 2.22
Weighted lead time according to particle swarm optimization and Tabu search algorithms at various weights.

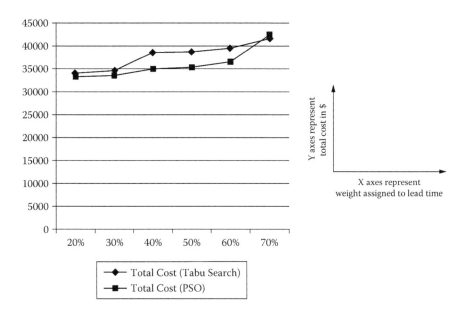

FIGURE 2.23
Total cost according to particle swarm optimization and Tabu search algorithms at various weights.

2.3.8 Solution to the Inventory Problem Using Multiobjective Genetic Algorithm

Now recalling Section 2.3.1, Equation (2.47), indicate total cost incurred by the supply chain including order processing, holding, and transportation costs in all participating facilities in the supply chain as follows:

$$TC_{(Total\ cost)} = TC_{(Retailers)} + TC_{(Warehouses)} + TC_{(Transportation)} \tag{2.47}$$

Based on the same section Equation (2.53) depicts weighted lead time of the supply chain as follows:

$$WLT_{(weighted\ lead\ time)} = \frac{\sum\limits_{j=1}^{no.\ of\ retailers} \dfrac{\sum\limits_{i=1}^{p} dp \times lp}{\sum\limits_{p=1}^{p} dp}}{no.\ of\ retailers} \tag{2.53}$$

Now, we have two conflicting objectives. The first is to minimize total supply chain cost (Equation 2.47). The second is to minimize weighted lead time (Equation 2.53). Therefore this problem comes under the category of a multiobjective problem. The multiobjectives problems constitute more than one conflicting objective and sets of constraints.

There are many efficient multiobjective evolutionary algorithms (MOEAs) that can be applicable to find out Pareto optimal solutions in the presence between two conflicting objectives: total supply chain cost and weighted lead time. The problem usually has not one unique solution, but a set of nondominated solutions, called the Pareto-optimal set (Carlos and Peter, 1998). The notion of Pareto-optimality is only a first step toward solving a multiobjective problem. Khor et al. (2005) discussed a conceptual framework for work to solve a multiobjective optimization problem where they have applied elitism strategy to ensure better search guidance.

The Nondominated Sorting Genetic Algorithm-2 (NSGA2) (Deb et al., 2002) is one of the most successful approaches in this regard. To solve this problem we have detained NSGA2 to obtain Pareto frontiers. There are three basic reasons for utilizing NSGA2 in solving this problem. First, it required less computational complexity of nondominated sorting. NSGA2 has maximum computational complexity and it has order of O (MN^2), where M is the number of objectives and N is the number of population sizes. In general almost all evolutionary algorithms have computational complexity, but it has almost no less than O (MN^3). Second, it incorporates a good deal of elitism. Due to this elitism approach, we also prevent the loss of good solutions. Third, there is an absence of any sharing parameters.

Constraints can be seen as high-priority objectives, which must be jointly satisfied before the optimization of the remaining, soft objectives takes place. Satisfying a number of violated inequality constraints is clearly the multiobjective problems of minimizing the associated functions until given values (goals) are reached. The concept of noninferiority is, therefore, readily applicable, and even particularly appropriate when constraints are themselves noncommensurable. When not all goals can be simultaneously met, a family of violating, noninferior points is the closest to a solution for the problem.

Nondominated fronts serve as a repository of a population of solutions and are used to compute successive generation by NSGA2. A set is defined as a nondominated set that is identified and contains a nondominated front of level 1 or front 1. To ensure the placing of individual solutions in the next nondominated front, it is the practice to discount the solution of front 1 and the procedure is iterated for the next solutions. Continue the above steps until we identify all the fronts. The idea of crowding distance to maintain the diversity in populations is adopted. Figure 2.24 describes the operational mechanism of NSGA2. Next, the algorithm applies binary tournament selection (to form the crossover pool), crossover, and mutation operators to generate the children population $C(1)$ of size L. Once initialized, the main part of the algorithm repeats for T generations. The algorithm applies nondominated sorting to $R(t)$, the resulting population from the union of parents $P(t)$ and children $C(t)$. The algorithm obtains the next generation population $P(t + 1)$ after selecting the L chromosomes from the first fronts of $R(t)$.

2.3.8.1 Calculation of Crowding Distance

The crowding distance of all the solutions in a nondominated set I has been calculated as follows:

Step A: Call the number of solutions in I as l, that is, $|I| = l$

Step B: First assign $I[i]_{distance} = 0$, for each i in the set.

Step C: Calculate the sorted indices vector: $I^m = sort(I, m)$ for each objective function.

Step D: Assign a large distance for boundary solution for $m = 1$ so that boundary points are always selected. That is, assign

$$I[1]_{distance} = I[l]_{distance} = \infty \tag{2.54}$$

Step E: Assign for other solutions, that is, for $i = 2$ to $(l - 1)$

$$I[i]_{distance} = I[i]_{distance} + \frac{\left(I[i+1].m - I[i-1].m\right)}{\left(f_m^{max} - f_m^{min}\right)} \tag{2.55}$$

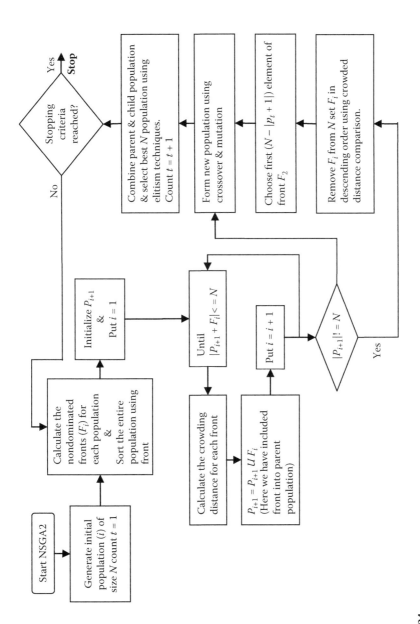

FIGURE 2.24
Flowchart of Nondominated Sorting Genetic Algorithm-2 (NSGA2).

Here, $f_m^{\max}$ and $f_m^{\min}$ are the population-maximum and population-minimum values of the mth objective function.

The present multiobjective optimization model is developed to search the large search space of possible solutions. The model is capable to reduce the large search space by finding nondominated solutions over successive generation of NSGA2. The NSGA2 result is very sensitive to algorithmic parameters; hence, it is necessary to perform several simulation runs to find suitable values for the parameters. The best parameters for NSGA2 have been selected through several test simulation runs as shown in Table 2.18. The best Pareto-optimal solutions (nondominated solutions) are obtained at 3,000 generations for a population size of 200. Each of these solutions represents an optimal subcontracting alternative for every stage of the network in the illustrative inventory management problem and provides an optimal trade-off in between total supply chain cost and weighted lead time.

Figure 2.25 shows the Pareto-optimal curve in between total cost and weighted lead time when we implemented NSGA2 in the previous inventory management numerical example.

TABLE 2.18

Best Nondominated Sorting Genetic Algorithm-2 (NSGA2) Parameters for Total Supply Chain Cost–Lead Time Trade-Off Analysis

NSGA2 Parameters	Parameter Value
Population size	200
Number of generation	3000
Crossover probability (P_C)	0.8
Mutation probability (P_m)	0.1

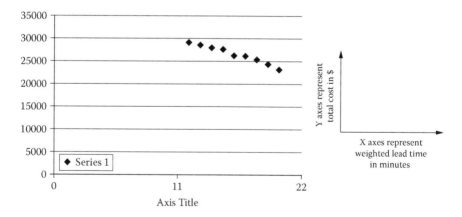

FIGURE 2.25
The Pareto-optimal curve between total cost and weighted lead time.

In this way we have shown the active role that management of inventory can play in making a supply chain effective and efficient. Here authors have tried to provide a clear-cut picture about implementation of both single-objective and multiobjective optimization algorithms for solving inventory management problems to make a supply chain more responsive.

2.4 Conclusions

In this chapter we discussed the responsive supply chain under the domain of network design and inventory management where the problems can be solved by employing metaheuristics optimization techniques. The discussion leads to responsive supply chain network design that percolates to the benefits of an integrated approach at the strategic decision-making level. In this network design problem we have tried to provide a feedback mechanism in between supply chain operation and planning to ensure the supplier's adaptation to a current market's demands. The model is formulated on the basis of multiechelon supply chain network design by considering the responsiveness as a design criterion. The network design problem identified the best set of suppliers, plants, distribution centers, and cross-docks to facilitate intended responsiveness.

The last part of the chapter shows how an efficient and effective inventory management policy not only regulates the cost incurred but also makes the supply chain responsive. Another salient feature of this chapter is the implementation of different types of metaheuristics optimization techniques to solve the network design and inventory management problem.

The suggested model can be extended to cover several other aspects of responsive supply chain, such as supplier selection, allocation of resources, and risk management. The model may be formulated under the domain of reverse supply chain.

References

Bachlaus, M., Pandey, M.K., Mahajan, C., Shankar, R., and Tiwari, M.K. 2008. Designing an integrated multi-echelon agile supply chain network: A hybrid taguchi-particle swarm optimization approach. *Journal of Intelligent Manufacturing*, 19, 747–761.

Carlos, M.F., and Peter, J.F. 1998. Multiobjective optimization and multiple constraint handling with evolutionary algorithms—part I: A unified formulation. *IEEE Transaction on Systems, Man, and, Cybernetics—Part A: Systems and Humans*, 28(1), 26–38.

Chao, I.-M., 2002. A tabu search method for the truck and trailer routing problem. *Computers and Operations Research*, 29, 33–51.

Chopra, S., and Meindl, P. 2001. *Supply Chain Management*. Upper Saddle River, NJ: Prentice Hall.

Deb, K., Amrit, P., Sameer, A., and Meyarivan, T. 2002. A fast and elitist multiobjective genetic algorithm: NSGA-II. *IEEE Transaction on Evolutionary Computation*, 6(2), 182–197.

Eglese, R.W. 1990. Simulated annealing: A tool for operations research. *European Journal of Operational Research*, 46, 271–281.

Ettl, M., Feigin, G.E., Lin, G.Y., and Yao, D.D. 2000. A supply network model with base-stock control and service requirements. *Operations Research*, 48, 216–232.

Glover, F. 1986. Future paths for integer programming and links to artificial intelligence. *Computers and Operations Research*, 13, 533–549.

Goldberg, D.E. 1989. Genetic Algorithms in Search, Optimization, and Machine Learning. Reading, MA: Addison-Wesley.

Golden, B.L., and Skiscim, C.C. 1986. Using simulated annealing to solve routing and location problems. *Naval Research Logistic*, 33, 261–279.

Hansen, P. 1986. The Steepest Ascent Mildest Descent Heuristic for Combinatorial Programming, Paper Presented at Congress on Numerical Methods in Combinatorial Optimization, Capri, Italy.

Huang, W., Romeijn, H.E., and Geunes, J. 2005. The continuous-time single-sourcing problem with production and inventory capacity constraints and expansion opportunities. *Naval Research Logistics*, 52, 193–211.

Ibaraki, T., and Katoh, N. 1988. *Resource Allocation Problems—Algorithmic Approaches*. Cambridge, MA: The MIT Press.

Johnson, D.S., Aragon, C.R., Mcgeoch, L.A., and Schevon, C. 1989. Optimization by simulated annealing: An experimental evaluation: Part I—graph portioning. *Operations Research*, 37, 865–892.

Khor, E.F., Tan, K.C., Lee, T.H., and Goh, C.K. 2005. A study on distribution preservation mechanism in evolutionary multi-objective optimization. *Artificial Intelligence Review*, 23, 31–56.

Kirkpatrick, S., Gelait, C.D.J.R., and Vechi, M.P. 1992. Optimization by simulated annealing. *Science*, 220, 671–680.

Kleijnen, J.P.C., and Smits, M.T. 2003. Performance metrics in supply chain management. *Journal of the Operational Research Society*, 54, 507–514.

Metropolis, N., Rosenbluth, A., Rosenbluth, M., Teller, A., and Teller, E. 1953. Equations of state calculation by fast computing machines. *Journal of Chemical Physics*, 21, 1087–1092.

Romeijn, H.E., Shu, J., and Teo, C.P. 2007. Designing two supply chain echelon networks. *European Journal of Operational Research*, 178, 449–462.

Ross, A., and Jayaraman, V. 2008. An evaluation of new heuristics for the location of cross-docks distribution centers in supply chain network design. *Computers and Industrial Engineering*, 55, 64–79.

Santoso, T., Ahmed, S., Goetschalckx, M., and Shapiro, A. 2005. A stochastic programming approach for supply chain network design under uncertainty. *European Journal of Operational Research*, 167, 96–115.

Sridhar, J., and Rajendran, C. 1993. Scheduling in a cellular manufacturing system: A simulated annealing approach. *International Journal of Production Research*, 31, 2927–2945.

Supply Chain Council. 2003. Supply chain operations reference model—Overview of SCOR Version 6.0. Supply Chain Council, Pittsburgh, PA.

Vilcot, G., and Billaut, J.C. 2008. A tabu search and a genetic algorithm for solving a bicriteria general job shop scheduling problem. *European Journal of Operational Research*, 190, 398–411.

Wilhelm, M.R., and Ward, T.L. 1987. Solving quadratic assignment problem by simulated annealing. *IEEE Transactions*, 19, 1, 107–109.

Yusuf, Y.Y., Gunasekaran, A., Adeleye, E.O., and Sivayoganathan, K. 2004. Agile supply chain capabilities: Determinants of competitive objectives. *European Journal of Operational Research*, 159, 379–392.

Yogeswaran, M., Ponnambalam, S.G., and Tiwari, M.K. 2009. An efficient hybrid evolutionary heuristic using genetic algorithm and simulated annealing algorithm to solve machine loading problem in FMS. *International Journal of Production Research*, 47, 19, 5421–5448.

3

Supply Chain Coordination: Modeling through Contracts

3.1 Introduction

A supply chain (SC) consists of a number of distinct entities that are responsible for converting raw materials into finished product and making them available to the final customers to satisfy their demand in the least possible time with the lowest possible cost. Managing the flows of materials, money, and information in the SC network implies the presence of many decision makers within the SC where each one operates a part of it. These decision makers could be either distinct firms or managers of different departments within a firm. When all the members of a SC are owned by a single individual, it is termed a *centralized supply chain* (CSC). On the other hand, when there are different owners for different entities, it is termed a decentralized supply chain (DSC). In today's business world, most of the supply chains encountered are decentralized in nature. Individual members of a DSC have conflicting objectives and when an individual member optimizes locally, it leads to an inefficient and high-cost solution for the entire SC. Because each player acts out of self-interest, we usually see inefficiencies in the system—that is, the results look different than if the system was managed optimally by a single decision maker who could decide on behalf of these players and enforce the type of behavior dictated by this globally (or centrally) optimal solution. Therefore, the major concern with supply chain management (SCM) is how to align the individual decisions with the entire SC objectives to arrive at a system optimal solution and reduce the SC inefficiency. Potential solutions to eliminate these inefficiencies are vertical integration, coordination through contracts, and collaboration.

3.1.1 Vertical Integration

In a vertically integrated firm, the entire control of the SC lies with a single central decision maker. Vertical integration allows a company to obtain raw materials at a low cost, and exert more control over the entire SC, in terms of both lead time and quality. An excellent example of vertical integration was Ford Motor Company early in the twentieth century (Swanson, 2003). In addition to automobile factories, Henry Ford owned a steel mill, a glass factory, a rubber tree plantation, an iron mine, and railroads and ships used for transportation. Ford's focus was on mass production, making the same car, Model T, cheaper and faster. This approach worked very well at the beginning and the price of the Model T fell from $850 in 1908 to $290 in 1924, and by 1914 Ford had a 48% share of the American market, and by 1920 it was producing half the cars made worldwide (Erhun and Keskinocak, 2011). But, in today's business environment, we do not have many examples of a vertically integrated firm except a very few such as the famous fashion design company ZARA. The main reason is that in today's fast-paced economy, the need and taste of the customer are changing very fast and the supply chain should be responsive. Therefore, the focuses of the companies are on their core competencies to stay ahead of their competitors to win their customers. Currently a trend has emerged toward virtual integration where SC is composed of independently managed but tightly linked companies.

3.1.2 Coordination through Incentives

Through incentive mechanisms, the objectives of individual members can be aligned with the SC objectives and can be coordinated. A procedure for aligning the plans of two or more decision-making units is called a coordination scheme (Stadtler, 2009). However, it is a challenging task to design incentive mechanisms or contracts such that even though each player acts out of self-interest, the decentralized solution might approach the centralized solution and thereby coordinates the SC. Such incentive mechanisms and contracts are useful if companies want to reduce the inefficiencies in their SC. SC coordination is a step toward SC collaboration, but it does not reap all the benefits of collaboration and also does not involve all the complications that arise due to collaboration in a SC.

3.1.3 Supply Chain Collaboration

Collaboration may be as simple as sharing information between different units or as involved as joint product design. Collaboration involves joint planning, joint product development, mutual exchange of information, integrated information system, cross-coordination on several levels in the companies in the network, long-term cooperation, and fare sharing of risks and benefits. SC activities with a high potential of benefitting from collaboration include new product introduction, procurement, logistics, replenishment planning, and demand forecasting. When the decision-making units are individual members of a SC, development of collaboration among the units is called *interfirm collaboration*, whereas collaboration between different divisions of the same organization is termed *intrafirm collaboration*. For example, by replacing the decentralized purchasing functions of a large company with centralized procurement, many companies were able to reduce the associated administrative costs, better utilize their procurement manager's time, and more importantly leverage their buying power with their suppliers, resulting in significant savings (Erhun and Keskinocak, 2011). This intrafirm collaboration is termed *horizontal collaboration* as the activities of the multiple units are responsible for the same function (i.e., purchasing). When collaboration takes place between different intrafirm functions such as production and distribution, production and sales, or production and procurement, it is termed *vertical collaboration*.

Interfirm collaboration takes place when independent companies work together, synchronize, and modify their business practices for mutual benefits, thereby shifting the nature of traditional relationships from adversarial to collaborative (Lapide, 1998). In interfirm collaboration, companies with similar characteristics, which are potentially competitors, collaborate on a particular business function (e.g., purchasing). On the other hand, collaboration between supplier and manufacturer or between distributor and retailer are examples of interfirm vertical collaboration.

It is interesting to note that though companies understand the benefits of collaboration, most of the time they are reluctant to change their SC practices. In that situation, it is desirable to design incentive mechanisms or contracts or change the term of the existing contract to align incentives and reduce inefficiencies in a SC. In this chapter, we categorize and review the substantive research on modeling SC coordination, and mainly it is expository in nature. We first discuss the nature of inefficiencies that might result from the decentralized decision making and how one can design contracts (coordination mechanisms) such that even though each player acts out of self-interest, the decentralized solution might approach the centralized optimal solution. This particular area has received attention from several disciplines such as economics, marketing, and operations management, and the ultimate motivation of everyone is to improve system performance and provide benefits to the final customer. For reviews on SC contracts and coordination, one can refer to Tsay et al. (1999) and Cachon (2001). This chapter will focus on specific incentive mechanisms to coordinate a SC mainly from the operations management perspective and will describe different analytical models.

3.2 What Is Coordination?

Here, we will look at a simple, stylized two-stage supply chain with one supplier and one retailer to understand SC coordination where the retailer purchases goods from the supplier and sells them in the end market (see Figure 3.1).

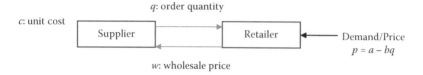

FIGURE 3.1
A two-stage supply chain.

For simplicity, assume that the supplier is uncapacitated and has unit cost of production c. The retailer faces a market where the price is inversely related to the quantity sold. For simplicity, let us assume a linear demand curve $P = a - bq$, where P is the market price and q is the quantity sold by the retailer. Further, it is assumed that all this information is common knowledge.

In a decentralized SC, the supplier and the retailer are two independently owned and managed firms, where each party is trying to maximize his or her own profits. The supplier chooses the unit wholesale price w and after observing w, the retailer chooses the order quantity q. Note that this is a dynamic game of complete information with two players, supplier and retailer, where the supplier moves first and the retailer moves second. The supplier's and the retailer's profits are $\Pi_S = (w - c)q$ and $\Pi_R = (a - bq - w) \, q$, respectively. Based on the wholesale price w, the retailer maximizes his or her profit by ordering quantity $q(w) = (a - w)/2b$ as the objective function Π_R is concave with respect to q. Based on this retailer optimum order quantity, the supplier decides his or her wholesale price. Substituting the value of optimum order quantity in the supplier's profit function and differentiating with respect to wholesale price w, one can determine the optimum wholesale price $w = a + c/2$. The profits of the supplier and the retailer are $\Pi_S = (w - c) q = (a - c)^2/8b$ and $\Pi_R = (a - c)^2/16b$, respectively. The optimal profit of the decentralized SC $\Pi = \Pi_S + \Pi_R = 3(a - c)^2/16b$.

On the other hand, in centralized SC (i.e., in a vertically integrated firm) a single entity determines the profit of the entire SC and $\Pi = (p - c) \, q = (a - bq - c)q$. The objective is to maximize the profit and differentiate the objective function with respect to q. The optimum order quantity for the centralized SC is $q = (a - c)/2b$ and the corresponding centralized SC profit is given as $\Pi = (a - c)^2/4b$.

Looking at the above equations, it is clear that SC profit under a centralized system is higher than the decentralized SC. In the DSC, the outcomes are worse for all the parties involved (supplier, retailer, supply chain, and consumer) compared to the CSC, because in the DSC both the retailer and the supplier independently try to maximize their own profits (i.e., each one try to get a margin, $P - w$ and $w - c$, respectively). This effect is called *double marginalization* (DM). Spengler (1950) provides early documentation with his findings on double marginalization where the retailer does not consider the supplier's profit margin when making order quantity decisions, and therefore orders too little quantity for system optimization. In a serial supply chain with multiple firms, there is coordination failure because each firm charges a margin and neither firm considers the entire supply chain's margin when making a decision. In this model, the profit loss in the DSC due to DM is 25%. For the sake of a firm, it is important to eliminate or reduce DM.

This simple model suggests that vertical integration could be one possible way of eliminating double marginalization. However, in today's business scenario, vertical integration is usually not desirable, or not practical as discussed at the beginning of the chapter. Then the question is, can we change the terms of the trade so that independently managed companies act as if they are vertically integrated? This is the concept known as supply chain coordination. In this stylized model, the retailer should choose optimal order quantity $q^* = (a - c)/2b$ in any coordinating contract. Accordingly, finding ways to provide incentives to members is an important issue. One can easily think of some very simple alternative contracts to eliminate DM and improve the performance of the DSC (see Erhun and Keskinocak, 2011).

3.3 Research on Coordination through Contracts

In this section, we will discuss different types of analytical models developed in the literature to coordinate a SC through various incentive mechanisms.

3.3.1 Price-Only/Wholesale Price Contract

The simplest of all contract types is the price-only (PO) contract. This is also known as a wholesale price (WP) contract and most of the studies are made under the newsvendor framework. Here, an upstream firm member sells an item to a downstream member as much as the latter wants, at a posted price. The retailer resells the goods to customers and keeps all the realized revenue, and the retailer is solely responsible for salvaging any unsold goods. The retailer will have to bear the risks of overstocking and understocking. To define the profit function of individual entities of SC, it is important to know the flow of payment, and in Figure 3.2, we have shown it under PO contract.

The manufacturer produces the product incurring a manufacturing cost m per unit and offers the product to the retailer at a wholesale price w per unit. In response to the offer made by the manufacturer, the retailer determines order quantity Q. The retailer sells these products to the end customer at a price p per unit. No returns are permitted in case of any leftover items at the end of the selling season. Therefore, the manufacturer pays an amount mQ toward the production cost and receives an amount of wQ from the retailer due to wholesaling. Because D is the random demand, the expected number

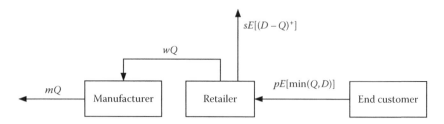

FIGURE 3.2
Flow of payments in a simple supply chain under price-only contract.

of sales at the retailer end can be given by $E\left[\min\left(Q,D\right)\right]$. This means that the retailer receives an amount of $pE\left[\min\left(Q,D\right)\right]$ from the customers due to the proceeds of the sale as shown in Figure 3.2. If s is the shortage cost per unit, then an amount of $sE\left[(D-Q)^{+}\right]$ is incurred by the retailer due to the unfilled demand of the customer, and it is shown in the figure as outflow of payment from the retailer. One can see the details in Parthasarathy (2011).

Mathematically, the retailer's expected profit equation is given by

$$E(\pi_R\left(Q\right)) = pE\left(\min\left(Q,D\right)\right) - w\,Q - sE\left(D-Q\right)^{+}$$

The retailer's objective is to maximize his or her profit function and determine the optimum order quantity Q^{*}. Based on the retailer's order quantity, the profit of the manufacturer is given as

$$\Pi_m = (w-m)Q^{*}$$

and the total profit of the SC is

$$\Pi_T = pE\left[\min\left(Q^{*},D\right)\right] - sE\left[\left(D-Q^{*}\right)^{+}\right] - mQ^{*}$$

Lariviere and Porteus (2001) have examined the effectiveness of PO contract using the newsvendor framework and analytically demonstrated that this type of contract cannot coordinate the supply chain. Even then considering the complexity of supply chains, the PO contract receives popularity because of its simplicity and serves as the benchmark in the analysis of supply chain performance and supply chain coordination (Tilson, 2008).

3.3.2 Quantity Discount Contract

Quantity discount (QD) is considered to be one of the most popular incentive mechanisms to coordinate a SC, and a lot of studies have been made on QD contract. In QD contract, the supplier induces the retailer to adopt the globally optimal order quantity, effectively compensating the retailer for deviating from his or her local optimal order quantity by providing incentive in the form of QD. Different types of discount mechanisms discussed in the literature are as follows:

- All-units quantity discount, where the lower price applies to all the units purchased
- Incremental quantity discount, where only those units within a price break interval receive the discount of that particular interval
- Business volume discount, where the price breakpoints are based on the total dollar volume of business across all products purchased from the vendor
- Time aggregation, where discount policy is applied to purchases within a given time frame (e.g., 1 year)

3.3.2.1 Coordination Models under Quantity Discount (QD) Contract

Here we have discussed models that have used QD as a coordination mechanism mainly under a deterministic environment. One can refer to the earlier work of Sarmah et al. (2006) for detailed discussion. Here, the words *vendor*, *supplier*, and *manufacturer* are used alternatively to represent the same upstream member in the SC who sells the item to the downstream member buyer unless specifically mentioned. In many practical situations, it is found that the manufacturer has a high production setup cost or high shipping cost. In such situations, the manufacturer prefers to produce or ship the product in large quantities. The manufacturer encourages buyers to order in large quantities by offering some incentives. Similarly, manufacturer also by ordering larger quantity to its raw material supplier (upstream member of the SC) may ask for some incentives from his supplier. Larger orders will lead to fewer productions setup and it may allow the vendor to take the advantage of economy of scale in transportation. Thus, it helps in improving the overall efficiency of the channel. The notations used in the model are as follows:

D = the buyer's annual demand for the product

S_i = Setup and ordering cost for the firm i, $i = \{1,2\}$

r_i = Annual inventory holding cost expressed as a percentage of the value of the item for the firm i, $i = \{1,2\}$ where subscript 1 and 2 represent buyer and vendor, respectively

Q = the buyer's order quantity

M_2 = the vendor/manufacturer's gross profit on sales expressed as a percentage

d_k = discount per unit offered by the manufacturer

R_2 = the manufacturer's production rate in units per year

P_0 = the buyer's base purchase price without quantity discount

C_2 = the manufacturer's manufacturing cost per unit excluding order processing, setup, and inventory holding costs per unit

3.3.2.2 Background

Traditional inventory models neglect the possibility of cost reduction by finding an order quantity that is jointly optimal for both the buyer and the vendor. With joint optimal policy the channel cost is reduced and the channel partners can negotiate to divide the savings (Thomas and Griffin, 1996). This saving opportunity by coordination of two entities of the system draws the attention of several researchers, and study of integrated inventory models can be viewed as one of the origins of SC coordination study from an operations management perspective. These models mainly examine the benefits accrued in the system due to coordination in order quantities between the two parties. Earlier, Goyal and Gupta (1989) reviewed the literature of buyer vendor coordination models. Benton and Park (1996) and Munson and Rosenblatt (1998) have also reviewed some of the papers discussed here under a different context. Sarmah et al. (2006) discussed QD contract in deterministic environment. We have mainly considered here the literature of channel coordination/SC coordination models that have an operations approach, and that literature mainly concentrates on the operating cost of the channel. Operating cost is considered as a function of retailer's/buyer's order quantity where a fixed retail price is assumed, and this leads to a fixed final demand.

The traditional inventory model assumes that a rational buyer would prefer to purchase his or her optimal order quantity (EOQ), as any deviation from this quantity would increase his or her total cost. The buyer's annual total cost for order quantity Q can be expressed as

$$TC(Q) = P_0 D + \left(\frac{D}{Q}\right) S_1 + \left(\frac{Q}{2}\right) r_1 P_0 \tag{3.1}$$

When quantity discount is not allowed, the buyer's optimal order size is given as

$$Q^* = \sqrt{\frac{2 D S_1}{r_1 P_0}} .$$

Thus, the total annual cost is given as

$$TC(Q^*) = P_0 D + \sqrt{2 D S_1 r_1 P_0} \tag{3.2}$$

Corresponding to buyer's order quantity Q, the vendor's/manufacturer's yearly net profit considering only order processing and setup costs for lot policy can be written as

$$YNP_2 = DM_2P_0 - \left(\frac{DS_2}{Q}\right) \tag{3.3}$$

The total channel cost is the sum of the individual cost component of buyer and vendor, respectively, and can be writen as

$$TCC(Q) = P_0D + \frac{D}{Q}(S_1 + S_2) + \left(\frac{Q}{2}\right)r_1P_0 \tag{3.4}$$

Vendor's/manufacturer's order processing and setup cost per order are considered to be larger than the buyer's order processing cost per order. If a buyer adopts his or her EOQ as the order quantity for minimizing his or her total annual costs, the vendor/manufacturer incurs a significant cost penalty. Therefore, the vendor/manufacturer induces the buyer through quantity discounts to order larger quantity to maximize profits. The manufacturer can maximize his or her profit when the lot size is as high as infinity. When a buyer purchases in larger order quantity $Q >$ EOQ, then there is an increase in profit for the vendor because of potential savings in order processing costs, manufacturing setup costs, and transportation costs. By selling fewer but larger orders, the vendor generates lower sales cost. Also, the vendor may save by seeking quantity discounts on raw materials the vendor receives from his or her supplier. The increased size of order quantity or lot size ultimately helps in improving the channel profits. An increase in profits should be shared in some equitable fashion so that coordination in a real sense is useful, and parties in the channel show interest to coordinate.Here, we categorized the various coordination models under QD as follows:

1. One can maximize the supplier's yearly net profit as shown by Equation (3.3) in our general model by adopting different lot size by giving an incentive to the buyer. The authors who have attempted the coordination problem from this perspective are classified here as vendor's/manufacturer's perspective coordination models.

2. Similarly, one can minimize the total system cost with respect to coordinated lot size or the order quantity as shown by Equation (3.4) and thereby improve the system savings. We have classified here those models as joint buyer and seller/manufacturer perspective coordination models.

3. On the other hand, some authors have studied the buyer/vendor coordination through quantity discount as a noncooperative and cooperative game. In a noncooperative game, each member will

try to maximize his or her profit or minimize his or her cost. Thus, the objective here will be to maximize Equation (3.3) and minimize Equation (3.1) of the general model. However, in a cooperative game, the objective will be to maximize system profit subject to the constraint that no player loses or incurs more from their noncooperative solution. We have categorized these models as a buyer and a seller/manufacturer coordination models under a game theoretic framework.

3.3.2.3 Coordination Models from Supplier's Perspective

Monahan (1984) in his model suggested that a vendor could encourage his customer to increase the order quantities from EOQ by offering a price discount. With the quantity discount, the buyer will be motivated to increase the order size up to KQ^* where K is a factor by which the vendor entices the buyer's order size. The amount of discount offered by the vendor compensates the buyer's increase in inventory costs. For the increased order size, the total cost of the buyer is given as

$$TC(KQ^*) = P_1 D + \sqrt{2DS_1 r_1 P_0}\left(1 + \frac{(K-1)^2}{2K}\right) \tag{3.5}$$

The increase in cost resulting from larger order size is the difference between the costs at the EOQ and costs at the order size KO^* as given by Equation (3.5). The vendor offers a price discount per unit equal to the increase in cost at the buyer's side, which is given as

$$d_k = \sqrt{\frac{2S_1 r_1 P_0}{D}}\left\{\frac{(K-1)^2}{2K}\right\} \tag{3.6}$$

Supplier's yearly net profit after giving a discount amount is given as follows:

$$YNP_2 = D(M_2 P_0 - d_k) - \frac{D}{KQ^*} S_2 \tag{3.7}$$

Substituting the value of d_k in Equation (3.7), maximize the supplier's profit equation YNP_2 with respect to K. The optimal value of K is obtained as

$$K^* = \sqrt{\frac{S_2}{S_1} + 1} \tag{3.8}$$

From the expression of K^* in Equation (3.8), one can easily say that when the value of S_2 is large, the supplier can entice the buyer to order in larger quantity, and the value of K^* is independent of the amount of discount offered by the supplier. One important issue here is that when the buyer is compensated for an increase in cost due to larger order size, the buyer will be indifferent to increasing his or her order quantity. Monahan developed the model considering lot-for-lot policy, an all-unit quantity discount schedule with single price break

3.3.2.4 *Joint Buyer and Seller Perspective Coordination Models*

Some authors have used quantity discount as a coordination mechanism to maximize the joint profit of the buyer and the vendor. The objective function here in all likelihood is to minimize the total channel cost as shown by Equation (3.4). The models here provide some explicit mechanism for division of surplus generated in the channel due to coordination. Like the seller's perspective model, here also it is assumed that the seller has complete information about the buyer's cost structure.

The idea of joint optimization for buyer and vendor was initiated by Goyal (1976) and later reinforced by Banerjee (1986). The objective of Goyal's model was to minimize total relevant cost for both the vendor and the buyer for the order quantity Q. He assumed that the manufacturer does not produce the item and in fact purchases it from another supplier. Moreover, he assumed that inventory holding costs are independent of the price of the item. Banerjee (1986) formulated a joint economic lot size (JELS) model for a buyer and a vendor system where the vendor has a finite production rate. He determines the JELS Q^* by differentiating the total system cost equation with respect to Q.

$$TC(Q) = \frac{D}{Q}(S_1 + S_2) + \frac{Q}{2}r\left(P_0 + \frac{D}{R_2}C_2\right) \tag{3.9}$$

$$Q^* = \sqrt{\frac{2D(S_1 + S_2)}{r\left(P_0 + \frac{D}{R_2}C_2\right)}} \tag{3.10}$$

The assumption they consider is that a production setup is incurred every time an order is placed. He finds that without quantity discount, the buyer incurs loss, but the supplier gets benefit if JELS is adopted rather than the buyer's EOQ. He developed the two bounds of discounts that allow the joint benefit to both parties if the buyer increases the order quantity from EOQ to the JELS quantity. When the discount amount is fixed at a lower bound, all the benefits go to the supplier and the buyer is indifferent, whereas when the amount of the discount is set at a maximum level, all benefits shift to the buyer and the supplier is indifferent. While suggesting equal distribution of the gains from joint economic ordering, Banerjee (1986) mentioned that

questions of pricing and lot-sizing decisions are settled through negotiations between the buyer and the seller. Later, we will see how some authors
have incorporated in their model the bargaining power of the channel members in fixing the order quantity and amount of discounts.

Viswanathan (1998) in his paper has compared two supply policies for an
integrated vendor/buyer inventory model. In the first policy, the vendor produces a batch and supply to the buyer in number of equal shipment size
at a constant interval. The second policy is to supply the production batch
to the buyer in increasing shipment size. He identified problem parameters
under which the equal shipment size policy and increasing shipment size
policy are optimal. The author has observed that neither of the two policies
dominates the other for all problem parameters. The second policy attempts
to shift inventory to the buyer as quickly as possible. This type of strategy
works better if the holding cost for the buyer is not much higher than that for
the vendor. Most of the models discussed above consist of two stages of SC.
Some authors have extended the two-stage model into a multistage model
and studied the coordination issues.

Zhou and Wang (2007) developed a general production-inventory model
for a single-vendor–single-buyer integrated system relaxing the assumption
of earlier models that buyer's unit holding cost should be higher than that
of vendor's holding cost. They also considered shortages occurring at the
buyer's end only. The model was extended to account for deteriorating items.
They identified three significant insights. First, no matter whether the buyer's
unit holding cost is greater than the vendor's or not, they claimed that their
model always performs best in reducing the average total cost as compared
to the existing models. Second, when the buyer's unit holding cost is less
than that of the vendor's, the optimal shipment policy for the integrated system will only include shipments increasing by a fixed factor for each successive shipment. In the following section, we will discuss the SC coordination
model with more stages. Ben-Daya and Al-Nassar (2008) have dealt with
inventory and production coordination in a three-layer supply chain involving suppliers, manufacturers, and retailers. In their model, it is assumed that
the cycle time at each stage is an integer multiple of the cycle time of the
adjacent downstream stage. The cost minimization model is found to impart
a considerable amount of coordination benefits in the channel.

3.3.2.4.1 Extension of the Model to Three-Stage SC

A two-stage SC is the basic building block of any complex SC, and many
authors have extended the basic model to more than two stages to capture the
real situation. Munson and Rosenblatt (2001) have extended the two-level SC
to a three-level SC by considering a supplier (who is supplying raw materials
to manufacturer), a manufacturer, and a retailer, and they explored the benefit of using a quantity discount on both ends of the supply chain to decrease
cost. Like the earlier scenario, the manufacturer's production lot size is an
integer multiple of the buyer's order quantity, and the manufacturer orders

an integer multiple of his production lot size to the raw materials supplier. They have shown that by quantity discount mechanism, a company can coordinate its purchasing and production functions. This creates an integrated plan that dictates order and production quantities throughout a three-firm channel. They have considered manufacture as the dominant member in the channel who takes the lead role in coordinating the channel.

Khouja (2003) has also considered a three-stage supply chain of a tree-like inventory model structure. He has considered three coordination mechanisms between the members of the supply chain and has shown that some of the coordination mechanisms can lead to a significant reduction in total cost. The author, however, has not considered the distribution of savings between the different members of the supply chain. Khouja (2003) also studied coordination of the entire supply chain from raw materials to customer considering single and multiple components. The author has considered components scheduling decisions at each stage in which manufacturing occurs and its impact on the holding cost. He has shown that complete synchronization in the chain leads to loss of some members of the supply chain and has provided an algorithm for optimal synchronization and incentive alignment along the supply chain.

Lee (2005) considered a vendor, a buyer inventory control problem where the manufacturer orders raw materials from the supplier and converts them into the finished products and finally delivers the finished product to the buyer. They have considered six costs, namely raw material ordering cost and holding cost, manufacturer's production setup cost and its finished goods holding cost, and buyer's inventory ordering cost and holding cost. Lee's objective was to develop an economic lot size model to minimize the integrated supply chain costs, while simultaneously taking all the above six costs into account. Lee and Moon (2006) developed inventory models for the three-level supply chain (i.e., a supplier, a vendor [warehouse], and a buyer [retailer]). The focus of their problem was determining the optimal integer multiple of time interval between successive setups and orders in the coordinated inventory model.

Jaber et al. (2006b) studied a three-level SC model with a profit-sharing mechanism to maximize the SC profit. The authors have used an all-unit price discounts scheme to coordinate the order quantities among the different members of the SC, and the demand at the retailer's end is assumed to be price dependent. To enhance coordination, two profit-sharing scenarios are investigated. The semiliberal scenario is based on increasing the quantity discount in order to generate more demands with which the most powerful player in the chain will get the highest fraction of additional profits. However, a strict mechanism is suggested to rectify the first scenario by dividing coordination profits based on equal return on investments. Banerjee et al. (2007) developed an integrated inventory model for coordinating the procurement of input materials, albeit in somewhat of a limited way, with the production schedule, which, in turn, is linked to the product distribution and delivery plan. They adopted the concept of integer lot size factors as potentially effective mechanisms for establishing linkages among inventories at various echelons of the supply chain for achieving coordination.

In recent times, few researchers have extended the model to four-stage SC in the same logic as discussed in the previous models. Pourakbar et al. (2007) considered an integrated four-stage SC incorporating one supplier, multiple producers, multiple distributors, and multiple retailers. The aim of this model is to determine the order quantity of each stage (from its upstream) and shortage level of each stage (for its downstream) such that the total cost of the supply chain is minimized. Their model is an extension of the earlier work of Lee (2005) and authors developed a heuristic solution procedure based on the genetic algorithm to solve this problem. Cárdenas-Barrón (2007) extended the work of Khouja's (2003b) three-stage supply chain by presenting an n-stage multicustomer supply chain inventory model and solved the cost function algebraically. They have limited their mathematics and numerical examples to a four-level supply chain considering a single product.

3.3.2.5 Coordination Models under Game Theoretic Framework

Some authors have viewed the buyer/vendor coordination problem through the quantity discount mechanism as a two-person game. They have formulated it as a nonzero sum game having elements of both conflict and cooperation. In a noncooperative game playing independently, the intention of the players (vendor and the buyer) is to maximize their individual gain. The objective function for this game from the general model can be written as

$$\text{Minimize } TC = P_0 D + \frac{DS_1}{Q} + \frac{Q}{2} r_1 P_0 \tag{3.11}$$

$$\text{Maximize } YNP_2 = DM_2 P_0 - \frac{DS_2}{Q} \tag{3.12}$$

Generally, the solution to the noncooperative game can be obtained by using an established equilibrium concept. Different types of game models have different solutions. In the Stackelberg game, the player who holds the more powerful position is called the leader and enforces his or her strategy on the other player. The other player who reacts to the leader's decision is called the follower. The solution obtained to this game is the Stackelberg equilibrium solution. In a cooperative game, both buyer and seller would consider that maximizing system profit subject to the buyer's total annual cost at cooperation should be less than or at most equal to those at noncooperation. Similarly, a seller's total annual profit at cooperation should be greater than, or at least equal to those at noncooperation. The objective function for this game from the general model can be written as

$$\text{Max } \lambda YNP_2 - (1 - \lambda)TC \tag{3.13}$$

$$\text{Subject to } TC \leq TC^*$$
$$YNP_2 \geq YNP_2^*$$

where TC^* and YNP_2^* represent the cost and profit of buyer and seller before cooperation. Depending upon the bargaining power of the seller and the buyer, the value of λ varies between 0 and 1. In the cooperative game a group of strategies is called a *Pareto efficient point* when at least one player will be better off and no player will be worse off from the initial condition. In the decentralized supply chain where the members belong to two different firms, the method of bargaining and negotiated solution that is dynamic in nature may result in better coordination in the supply chain as compared to a static coordinated solution in a centralized supply chain.

Chiang et al. (1994) studied the quantity discount problem from the game theory perspective considering both cooperative and noncooperative game models. They have shown the benefits resulting from cooperation between the buyer and the seller. Christy and Grout (1994) prescribed a model to safeguard the relationship between buyer and supplier in a supply chain using the principles of game theory and transaction cost economics.

Li et al. (1995, 1996) also studied buyer/seller cooperation assuming the buyer is in a monopolistic market for the product in a constant demand situation under a game theory framework. Comparing the cooperative and noncooperative models, they showed that system profit is higher at cooperation than at noncooperation, and the wholesale price of the seller to the buyer is lower at cooperation than at noncooperation. Further, authors have shown that the quantity discount approach is an effective mechanism for achieving system cooperation. Cachon and Zipkin (1999) studied a serial supply chain with one supplier and one buyer where demand is stochastic under a game theoretic framework. Considering local and echelon inventory, they framed two games: the echelon inventory game and the local inventory game. They found that in each game there is usually a unique Nash equilibrium. They also determined a system optimal solution where the firm chooses base stock policies. Because the optimal solution is not Nash equilibrium, the authors have observed that competitive decision making degrades supply chain efficiency and have evaluated the magnitude of this effect with an extensive numerical study. Leng and Parlar (2005) provided a review on game theoretic model on SCM. Esmaeili et al. (2009) studied a single buyer and single supplier coordination problem from the game theoretic point of view and modeled the buyer and supplier relationship by cooperative and noncooperative game. They incorporated competition and cooperation between the two parties of the SC.

Ertogral and Wu (2001) have raised the issue that SC contracting should also consider the aspect of negotiation between the two parties for final settlement of the division of surplus generated due to supply chain coordination.

They examined a bargaining theoretic approach for splitting total surplus generated in the system due to supply chain coordination between the two parties. The model allows for predicting the negotiation outcome between a buyer and a supplier considering their outside options and probability of breakdown of negotiation. Chu and Lee (2006) studied an information sharing problem in a vertical supply chain with one vendor and one retailer and employed the perfect Bayesian equilibrium as the solution concept used in a dynamic game played under incomplete information. Costantino and Di Gravio (2009) combined the concepts from game theory and fuzzy logic to analyze a bargaining problem with incomplete information.

From the available literature, it is noted that application of both cooperative and noncooperative game theories to SC coordination is at its infancy, and lot of scope is there to make the theories and their application popular among researchers and practitioners.

3.3.2.6 Coordination Models between Single Seller and Multiple Buyers

Many authors have studied the coordination issue for single manufacturer and multiple buyers' cases. To differentiate single manufacturer, multiple buyer's literature is a difficult task as in many cases, many authors after developing the model for the single buyer case have extended it for multiple buyers' cases, and these models are mostly confined to a homogeneous group of buyers. In one of the earliest works, Lal and Staelin (1984) initially studying one vendor and one group of homogeneous buyers have extended the study to a heterogeneous group of buyers for determination of optimal pricing policy. His heterogeneous groups differ in size, holding cost, order cost, and demand rates. They vary between the groups but not within a group. They have not obtained a close form solution. The solution to the heterogeneous group of buyers is offered by considering continuous approximation of a discrete quantity discount schedule.

Banerjee and Burton (1994) in their study on the multiple buyers' case, assuming a deterministic situation and vendor's demand rate as approximately constant, have shown that in multiple buyers' case, the classical economic lot size model may not be able to truly reflect the exact scenario due to discrete vendor inventory depletion. They observed that even under a deterministic situation, in the absence of an adequate production reorder point policy, stock-out may occur. They considered a common replenishment cycle–based coordinated inventory model and showed that it is superior to the individual optimization approach in the multiple buyers' case. Further, they extended the above model for stochastic demand and lead time conditions and proposed an iterative algorithm for solving the model.

Ingene and Parry (1995) investigated a two-level vertical channel consisting of a manufacturer and multiple independent retailers from a marketing point of view. They have shown that in a situation where the retailers have common territory, a two-part tariff wholesale pricing policy will fully coordinate the channel. Further, they have shown that the optimum number of retailers for channel coordination for profit maximization is determined through a fixed fee component of a two-part tariff. Lu (1995) considered an integrated inventory model with a vendor and multiple buyers. He assumed the case where the vendor minimizes its total annual cost subject to the maximum cost that the buyer may be prepared to incur and formulated it as a mixed integer programming problem. The author considered quantity discount schedules to maximize the vendor's total profit subject to the maximum cost that the buyer may be prepared to incur.

Gurnani (2001) also studied quantity discount pricing models with different ordering structure in a system consisting of a single supplier and heterogeneous buyers. This work is an extension of the earlier work of Zahir and Sarkar (1991) where authors considered a price-dependent demand function for multiple regional wholesalers who are served by a single manufacturer. Gurnani (2001) considered order coordination, order consolidation, and multitier ordering hierarchy. He showed that for identical buyers, order coordination leads to a reduction in system cost. For heterogeneous buyers, they determined the sufficiency condition when the coordination will be preferable.

Viswanathan (2001) has shown that a vendor could implement the common replenishment period mechanism by offering price discounts to buyers in a one-vendor, multibuyer supply chain for a product with the assumption that the vendor follows lot-for-lot policy. Under the proposed strategy, the vendor specifies the common replenishment period for all the buyers, which is a cost-increasing alternative for the buyers. Therefore, the vendor offers a price discount to encourage the buyer to accept this strategy. The price discount must be such that it compensates buyers for any increase in inventory costs and possibly provides additional savings. The important assumption in the model is that the buyer's cost and demand parameters are known; therefore, the vendor can anticipate the buyer's reaction. Mishra (2004) extended the above model of Viswanathan (2001) by considering selective discount policy. He studied the common replenishment period mechanism by allowing some buyers to participate in the coordination scheme where they get a discount for ordering a larger quantity, whereas other buyers continue to order in the earlier fashion without going for a discount. They found that in some situations, it might be beneficial to segment the buyers by offering multiple common replenishment periods.

Woo et al. (2001) studied an integrated inventory system where a single vendor purchases and processes raw materials in order to deliver them to multiple retailers. With the objective that both vendor and buyers are willing to invest to reduce joint ordering cost, the authors developed an analytical model to derive the optimal investment amount and replenishment decisions for both vendor and buyers. They have shown that the vendor and all buyers can obtain benefit directly from cost savings.

Chen, Federgruen, and Zheng (2001) in their work of one supplier and many retailers have shown that the same optimum level of channel-wide profit of a centralized system can be achieved in a decentralized system only if coordination is achieved via periodically charged fixed fees and a nontraditional discount pricing scheme. Under such a scheme, the discount given to the retailer is the sum of the three discount components based on the retailer's annual sales volume, order quantity, and order frequency. Further, they have shown that they not only have a traditional discount scheme based on order quantities but also the capability to optimize channel-wide profit when there are multiple nonidentical retailers. Wang (2002) presented an analysis for a supplier's quantity discount decision for heterogeneous buyers. For a single buyer and a group of homogeneous buyers, a single price break is sufficient for the supplier to maximize his or her quantity discount gain, but for multiple heterogeneous buyers a single price break point is not sufficient. He analyzed a supplier's quantity discount decision by using a common discrete all-unit quantity discount schedule for all the buyers. Boyaci and Gallego (2002) analyzed coordination issues in a supply chain consisting of one wholesaler and one or more retailers under deterministic price-sensitive customer demand and focused on inventory and pricing policies that jointly maximize channel profit. They showed that an optimal policy can be implemented cooperatively by an inventory consignment agreement. Also, it is mentioned that the above policy is capable of distributing the gains of channel coordination without requiring side payments.

Wang and Wang (2005) studied a decentralized supply chain, in which a supplier sells a product to a group of independent retailers, and analyzed the supplier's optimal QD policy. They considered a setting in which retailers are located in geographically dispersed areas, with each facing a demand that is a decreasing function of its retail price. This type of setting represents a common situation in reality (e.g., manufacturers often sell their products to privately owned distributors/retailers in different areas). Siajadi et al. (2006) developed a methodology to obtain the joint optimum order quantity in the case where multiple buyers are ordering a single item from a single vendor. They determined the shipment policy such that total cost is minimum. Further, it is shown that a multiple shipment policy is more beneficial than a single shipment policy considered by Banerjee (1986). It is shown that saving is increasing as the total demand rate approaches the production rate (i.e., the better the production capacity utilization is, the greater the saving will be in the SC). Conversely, when the dominating cost is the transportation cost, the saving is decreasing as the numbers of shipment approach one. Consequently, the new model becomes identical with the traditional model, as the numbers of shipment are equal to one.

3.3.2.7 Some Insights and Limitations

From the study of the above models, it is seen that this stream of literature describes the SC in a highly aggregated level and often considers only two decision makers. The important insight provided by the above literature is that there is an increase in profit for the manufacturer when the buyer purchases more than his or her EOQ. It is based primarily on the facts that the manufacturer's setup cost is much higher compared to the buyer's ordering cost, and the manufacturer may use a production cycle that is an integer multiple of the ordering cycle of the buyer:

- Many models are developed considering the holding cost of the buyer to be independent of purchase price.

- With few exceptions, models are developed where the supplier offers all-unit quantity discount with a single-price break point. Further, the manufacturer is assumed to have two ways of acquiring the item either by outside purchasing or manufacturing the item subject to specific production capacity.

- Most of the models simplify the purchasing/production system to one product and one machine.

- Many of the models fail to specify how the incremental savings to the manufacturer can be passed onto the buyer. Some authors mentioned equal splitting of the surplus, whereas some suggested splitting the surplus according to their investment. Most of the models are silent about conflict resolution between the supply chain partners (e.g., division of surplus between buyers and suppliers). Such a problem may call for the use of a game theoretic negotiation model.

- Most of the models assume that a supply chain partner has complete information (including cost, demand, lead time, etc.) about the other partner. This is considered to be a major limitation of these models. In a decentralized supply chain, it will rarely be the situation where complete information will be available with the parties. Coordination under limited information sharing is an important issue of concern to be studied for the decentralized supply chain.

- Single-vendor, multiple buyers' literature is still in its infancy state. Particularly, how a supplier should develop a quantity discount schedule when dealing with many buyers with different demand and cost structure is not known. Thus, the mechanism for additional profit sharing between vendor and multiple heterogeneous buyers is an important issue that needs investigation.

- In single-vendor, multiple buyers' literature very little work is available considering the vendor as a manufacturer producing the items to supply multiple heterogeneous buyers. In such a situation, how to tackle the discrete vendor inventory depletion in the model is an area that requires further study. We have tried to discuss the various future research issues on SC coordination.

Another incentive mechanism similar to QD is the credit mechanism. Many authors have used this mechanism to coordinate a SC, and in the next section, we discuss the mechanism.

3.3.3 Coordination through Credit Mechanism

In the inventory management literature, it is normally assumed that payment will be made to the vendor for the goods immediately after receiving the consignment. However, it is found that many industries provide a specific credit period for payment after items are delivered to the buyers. Several researchers have developed analytical inventory models with the consideration of a permissible credit period for payments. In SC credit has been used as an incentive mechanism to coordinate a SC. Before reviewing the coordination model with credit option, we briefly discuss in the following section the importance of a credit option in the business environment.

3.3.3.1 Importance of Trade Credit Option in Today's Business Environment

The practice of the seller providing credit to its customers is not a new phenomenon. But in today's business environment, it has a new dimension and becomes an important form of financing for business. Particularly in developing countries like India, the role of trade credit is immense, where growth of the financial institution is less compared to in developed nations. Trade credit can be defined as the purchase of goods or services, which involves delivery of goods or services at a certain date with payment at a later date (Issakson, 2002). One can see the various motives linked with trade credit as discussed by authors such as Ellienhausen et al. (2003). Some of the important reasons trade credit is relevant to our study are as follows:

Trade credit as sales promotion motive:

- In today's customer-focused business, credit option plays a very important role, and it can be used as a tool to compete in the market for generating sales. Through credit, a supplier can gain a competitive edge over the competitors. The length of the credit period offered by the supplier can be considered as a strategy against another supplier for winning over the buyer's order.
- Credit extension helps in developing good, long-term relationships with customers, which ultimately helps in generating future income for the firm. Further, credit option can generate repeat purchase orders of the customers (Wilson et al., 2000).
- Sometimes to off-load some of his excess inventories into a buyer's space, the supplier is willing to provide credit to the customer.

- A manufacturer may differentiate their offering to the market by extending trade credit to their customers as a commitment to the quality of their products. Due to this credit option, the customer is getting time to inspect the quality of the product before paying for the goods.

Trade credit as transaction motive:

- In the absence of trade credit, firms must pay for purchases upon delivery. When the timing of delivery of goods is uncertain and converting liquid asset into cash is costly, most firms have to hold precautionary cash balances. The use of trade credit provides information on future cash needs by allowing buyers to accumulate invoices for payment. The information enables firms to better predict their cash needs. As a result, they are able to hold smaller cash balances and lower brokerage cost. Similarly, the supplier also benefited from credit as the supplier can also predict cash receipt more accurately.
- Further, a larger firm has better access to financial institutions as compared to its smaller counterparts. For running the business, many times a smaller firm has to depend on the offerings of the trade credit of the larger business firm. Often large firms provide credit to the smaller firms and offer cash discounts for early payment, which ultimately helps in the process of finally selling the product.

Moreover, in developing countries, there are number of factors that create uncertainty in the delivery of products. Some of the important ones are poor transportation infrastructure and lack of proper communication infrastructure facility. These factors compel an inventory-intensive firm to maintain a high level of inventory. In many situations, small firms do not have sufficient working capital to maintain inventory for such a long period. In such an environment, trade credit can help in improving the productive efficiency of the organization as well as efficiency of the supply chain as a whole. The literature on inventory models with permissible delay in payment is quite vast. Here, an attempt has been made to cover only those models that have studied from the SC coordination perspective. A simple model is discussed here to show the benefits of SC coordination achievable through credit option. For a more detailed discussion, one can refer to Sarmah et al. (2007).

3.3.3.2 Coordination Models with Credit

Use of credit period/delay in payment as incentive mechanism in the literature got much attention since the mid-1980s. Chapman et al. (1985) have shown that by considering delay in payment in the inventory control model, a significant amount of inventory cost can be reduced and also at the same time it can be used as an effective marketing strategy. The other contemporary work to be mentioned here is that of Kingsman (1983) and Davis and Gaither (1985). Goyal (1985) developed a mathematical model for obtaining the economic order quantity for an item for which the supplier permits a fixed delay in settling the amount owed to him. He used the average cost analysis approach to show that savings in cost as a result of permissible delay in settling the replenishment account largely comes from the ability to delay payment without paying any interest. During the delay period, the supplier actually gives a loan to the customer without interest so that the customer can sell the items, accumulate revenue, and earn interest.

Looking at SC coordination literature, it is found that in recent times, many authors have extended the previously studied inventory models in the context of SC and used credit as an incentive mechanism to coordinate a SC. Moses and Seshadri (2000) have used delay in payment option as an incentive mechanism for two-stage SC coordination. Boyaci and Gallego (2002) have compared trade credit and inventory consignment contract. They mentioned that consignment coordination is more desirable because with trade credit coordination the buyer's and supplier's capital costs to converge and therefore render the contract infeasible in the long run.

Jaber and Osman (2006a) proposed a credit period as a decision variable to coordinate the order quantity between a two-level SC with the basic assumption that the setup cost of a supplier is higher than the ordering cost of the buyer. Through numerical study, the authors showed that with coordination, the retailer orders in larger quantities than its EOQ and savings take place to either both players, or to one in the SC. The authors also developed a mechanism to distribute the surplus generated due to coordination.

Chen and Kang (2007) considered a similar model to that of Jaber and Osman (2006a), where they investigated their model for predetermined and extended periods of delay in payments. However, and unlike the work of Jaber and Osman (2006a), Chen and Kang (2007) have not treated the length of delay in payment as a decision variable. Sheen and Tsao (2007) consider vendor/buyer channels subject to trade credit and QD for freight cost. Their work determined the vendor's credit period, the buyer's retail price, and order quantity while still maximizing the profits. They focused on how channel coordination can be achieved using trade credit and how trade credit can be affected by QD for freight cost.

Sarmah et al. (2007) considered a coordination problem that involves a vendor (manufacturer) and a buyer where the target profits of both parties are known to each other. Considering a credit policy as a coordination mechanism between the two parties, the problem's objective was to divide the surplus equitably between the two parties. Further, they compared trade credit and QD contracts under the EOQ framework for a two-stage SC and showed that trade credit is a better incentive mechanism to coordinate a SC if the supplier's cost of capital is lower than the buyer's cost of capital. Otherwise a QD contract is more efficient. Further, Sarmah et al. (2008a) studied single manufacturer and multiple buyer SC coordination problems through credit option and determined the optimal credit period that entices buyers to accept coordinated shipment. In the following section, we discussed a simple model to show how coordination between two parties can be achieved through credit option and benefits accrued out of it.

3.3.3.3 Model for Coordination through Credit

In this section, how credit can be used as a coordination mechanism is presented through a model. Details of the model can be found in Sarmah et al. (2007).

   For fixed selling and procurement prices for a product, there is no incentive for a buyer to order an amount except his or her EOQ if the buyer's objective is to maximize profit. Similarly, when setup cost of the manufacturer is high, for a fixed selling price and manufacturing cost, the manufacturer maximizes his or her profit if his or her lot size is as high as possible (i.e., infinity). If manufacturer lot size is equal to the buyer's EOQ, the profit of the manufacturer is minimal. On the other hand, if the manufacturer's batch size is infinity, the profit of the buyer is minimal. Thus, if the manufacturer's batch quantity (Q_m) is higher than the buyer's EOQ (Q_b), the profit of the manufacturer increases whereas that of the buyer decreases. In such an environment, to do business together the batch quantity could be between economic order quantities of the buyer (Q_b) and ∞ provided the buyer gets profit higher than profit at her EOQ and the manufacturer also gets profit higher than his or her minimum. In fact, both may have their own target profit and both may want to see that their target profits are ensured.

3.3.3.3.1 Assumptions and Notation

In the development of the mathematical model all the basic assumptions of the EOQ model are valid. In addition to that it is assumed that the cost of borrowing for the buyer is the same as the return that buyer gets by lending. This assumption excludes the possibility of making more profit by lending. The notations are as follows:

   D = Annual demand

   S_i = Setup/ordering cost of entity i for each setup/order (where $i = m,b$ represents manufacturer and buyer, respectively)

   h_i = Holding cost of entity i per unit per year

   C = Unit cost of acquiring the item by the manufacturer

   R = Unit retail price of the buyer

   P = Contract price for each unit between the buyer and the manufacturer

   Q_j = Joint economic order quantity

   Q_b = Buyer's economic order quantity (EOQ_b)

   Q_m = Manufacturer's batch quantity

t = Credit time in year
TC_b = Total annual inventory related cost to the buyer
EBQ_m = Manufacturer's economic batch quantity
YNP_i = Yearly net profit of entity i

3.3.3.4 Mathematical Model of the Problem

We developed a two-stage SC coordination model with credit period as a coordination mechanism for the situation when the two parties of the SC are willing to share complete information including each other's target profit. Further, the focus of the model is to develop the coordination mechanism such that the two parties of the SC can divide the surplus generated due to coordination after satisfying their own target profits. We have considered here two cases. First, we considered the situation when the two parties have no predecided individual target profit. The case of predecided individual target profit is considered next.

Case I: No Predecided Individual Target Profit for the Two Parties

Here, both parties of the supply chain do not have any predecided target profit. The model assumes that the two parties are willing to adopt a common order quantity $Q > (EOQ_b)$. As a result of this, additional profit generated at the manufacturer's side is shared equally through a credit giving mechanism.

Initially, the buyer is ordering at her EOQ to the manufacturer. Total annual inventory related cost to the buyer when she orders at her EOQ is given as

TC_b = Ordering cost + Inventory carrying cost

$$TC_b = \frac{DS_b}{Q_b} + \frac{Q_b}{2} h_b, \text{ where } Q_b = \sqrt{\frac{2DS_b}{h_b}}$$

Substituting the value of Q_b in the above equation, we get

$$TC_b = \sqrt{2DS_b h_b} \qquad (3.14)$$

Profit of the manufacturer when buyer orders at her EOQ is given by

$$YNP_m = (P - C)D - \frac{DS_m}{Q_b} \qquad (3.15)$$

When the manufacturer's setup cost is higher than that of the buyer's ordering cost, the manufacturer will induce the buyer to order in quantities larger than her EOQ, and any increase in cost at the buyer's side is compensated for through credit. If $Q > Q_b$ is the ordering quantity of the buyer, then the buyer's total inventory related cost is given as

$$TC_{b1} = \frac{DS_b}{Q} + \frac{Q}{2} h_b \tag{3.16}$$

An increase in buyer's cost, Z, is the difference of the cost at the buyer's new ordering quantity Q and the cost at her EOQ, can be written as

$$Z = \frac{DS_b}{Q} + \frac{Q}{2} h_b - \sqrt{2DS_b h_b} \tag{3.17}$$

This increase in cost of the buyer is to be compensated for by the manufacturer. The manufacturer's yearly net profit (YNP_m) due to changed ordering policy is given as

YNP_m = Gross revenue – Setup cost – Amount of compensation to the buyer equal to the increase in cost of buyer

$$YNP_m = (P-C)D - \frac{DS_m}{Q} - \frac{DS_b}{Q} - \frac{Q}{2} h_b + \sqrt{2DS_b h_b}. \tag{3.18}$$

Differentiating the above equation with respect to Q, the optimum value of order quantity Q

$$Q^* = \sqrt{\frac{2D(S_b + S_m)}{h_b}} \tag{3.19}$$

Now, the manufacturer compensates any increase in buyer's cost due to changed ordering policy through credit. The buyer will be interested to order a lot size Q larger than her EOQ, provided her cost does not increase from her initial cost.

The minimum credit time required by the buyer, which compensates his or her increase in cost due to coordinated order quantity from Equation (3.17), is given as

$$\frac{DS_b}{Q^*} + \frac{Q^*}{2} h_b - \sqrt{2DS_b h_b} = Dt_{min} h_b \tag{3.20}$$

$$t_{min} = \frac{S_b}{Q^* h_b} + \frac{Q^*}{2D} - \sqrt{\frac{2S_b}{Dh_b}} \tag{3.21}$$

The manufacturer's maximum yearly net profit with minimum credit time for coordinated order quantity Q^* is given as

$$YNP_{max} = (P-C)D - \frac{DS_m}{Q^*} - Dt_{min} h_m \tag{3.22}$$

The manufacturer can give credit to the buyer as long as his or her cost with credit due to coordinated order quantity Q^* is not more than cost without credit (i.e., cost incurred when buyer is ordering at his or her EOQ). From the difference of Equations (3.22) and (3.15), one gets the following condition:

$$\frac{DS_m}{Q^*} + Dt_{max}h_m = \frac{DS_m}{Q_b}$$

The maximum credit time that manufacturer can give to the buyer is

$$t_{max} = \frac{S_m}{h_m}\left(\frac{1}{Q_b} - \frac{1}{Q^*}\right). \tag{3.23}$$

Thus, t_{min} and t_{max} are the two bounds of the credit time over which negotiation between the two parties can take place.

The additional profit available at the manufacturer's side is the difference of profit at new ordering policy to the profit when buyer orders at his or her EOQ. Thus, the difference of Equations (3.22) and (3.15) gives the maximum extra profit the manufacturer obtains due to the change of the ordering policy. We considered here that this surplus can be divided equitably between the two parties by giving extra credit to the buyer. Let this extra credit period equal Δt. Therefore, we can write

$$2\Delta tDh_b = S_mD\left(\frac{1}{Q_b} - \frac{1}{Q^*}\right) - Dt_{min}h_m \tag{3.24}$$

$$\Rightarrow \Delta t = \frac{S_m}{2h_b}\left(\frac{1}{Q_b} - \frac{1}{Q^*}\right) - \frac{t_{min}h_m}{2h_b}$$

Through the credit-giving mechanism, the manufacturer is not only compensating for the buyer's extra cost but also at the same time shares equitably the surplus generated due to coordination by giving credit for an extra period of time Δt. Thus, total credit time t becomes

$$t = t_{min} + \Delta t$$

After substituting the proper value and on simplification, one gets the following expression:

$$t = \left(\frac{S_b}{Q^*h_b} + \frac{Q^*}{2D} - \sqrt{\frac{2S_b}{Dh_b}}\right)\left(1 - \frac{h_m}{2h_b}\right) + \frac{S_m}{2h_b}\left(\frac{1}{Q_b} - \frac{1}{Q^*}\right) \tag{3.25}$$

The buyer's profit after sharing the surplus equitably is given as

$$YNP_b = (R-P)D - \frac{DS_b}{Q^*} - \frac{Q^*}{2}h_b + D(t_{\min} + \Delta t)h_b \qquad (3.26)$$

Similarly, the manufacturer's profit due to sharing of the surplus equitably is given as

$$YNP_m = (P-C)D - \frac{DS_m}{Q^*} - D(h_m t_{\min} + h_b \Delta t) \qquad (3.27)$$

Case II: With Predecided Target Profit for the Two Parties

Two different parties, namely, the buyer and the manufacturer of a supply chain, depending on their power structure in the channel, may have some target profits from the business. Their target profits may be different. In the development of the model, we have assumed that both parties know each other's target profit and no party may be willing to coordinate so long as it achieves its target profit. Therefore, for a coordinated solution both parties' target profit must be fulfilled. Let the buyer's target profit, T_b, be

$$T_b = (1+x)T_{b\min}, \text{ where } x \geq 0$$

Similarly, target profit of the manufacturer

$$T_m = (1+y)T_{m\min}, \text{ where } y \geq 0$$

For a centralized solution with the credit option, the profit obtained in the channel is highest and can be expressed as

$$CP_{\max} = (R-C)D + Dt(h_b - h_m) - \sqrt{2D(S_b + S_m)h_b} \qquad (3.28)$$

From the above equation, it is seen that centralized profit is an increasing function of credit time, and it will be maximum when $t = t_{\max}$.

When $T_b + T_m > CP_{\max}$, then no solution is possible unless the target profits of both the parties are lowered proportionately. The revised target profit of the buyer and the manufacturer under such a situation will be

$$T_b' = \left(\frac{T_b}{T_b + T_m}\right)CP_{\max}. \qquad (3.29)$$

$$T_m' = \left(\frac{T_m}{T_b + T_m}\right)CP_{\max} \qquad (3.30)$$

Again, when $T_b + T_m \leq CP_{\max}$, then solution is possible.

Therefore, when the sum of the target profit is less than the profit obtained due to centralized solution, extra profit left over after achieving each individual target profit is divided equally between the two parties. Thus, the new profit of the manufacturer with initial target profit T_m will be

$$T_{m1} = T_m + \frac{1}{2}\left[CP_{max} - (T_b + T_m)\right] \tag{3.31}$$

Similarly, the new profit of the buyer with initial target profit T_b of the buyer will be

$$T_{b1} = T_b + \frac{1}{2}\left[CP_{max} - (T_b + T_m)\right] \tag{3.32}$$

We can also say that when $T_{b1} = \lambda CP_{max}$, where λ is a fractional value between 0 and 1, credit time required for the buyer to achieve the profit will be

$$t = \frac{\lambda CP_{max} + \dfrac{DS_b}{Q^*} + \dfrac{Q^*}{2}h_b - (R - P)D}{Dh_b}. \tag{3.33}$$

Correspondingly, the manufacturer's profit will be equal to $(1 - \lambda)CP_{max}$.

With a small numerical example we have illustrated the model.

3.3.3.5 Numerical Example

The following data are considered to illustrate the developed model:

Annual demand $D = 50$ units, setup cost $S_m = 25$ Mu per setup, ordering cost $S_b = 1.5$ Mu per order, holding cost of buyer per unit per year $h_b = 1.5$ Mu, contract price between the buyer and manufacturer $P = 5$ Mu per unit

Assumed data: Holding cost of manufacturer per unit per year $h_m = 1.2$ Mu

Cost of acquiring the item by the manufacturer $C = 2$ Mu per unit, sell price of the item by the buyer $R = 6$ Mu per unit

Economic order quantity of the buyer $Q_b = 10$ units

Channel optimum order quantity $Q^* = 42$

Assumed value of target profit of buyer $T_b = 50$ Mu

Assumed value of target profit of manufacturer $T_m = 70$ Mu

Minimum credit time offered by manufacturer $t_{min} = 0.243$ year

Maximum credit time manufacturer can offer $t_{max} = 1.58$ year

TABLE 3.1

Details of Profit Distribution between the Members of the Channel

Channel Members	Without Coordination (Mu)	With Coordination and Equitable Distribution of Surplus			
		Generated in the Channel Due to Minimum Credit Given to the Buyer (CP_{min})		Generated in the Channel Due to Maximum Credit Given to the Buyer (CP_{max})	
		With Target Profit (Mu)	Without Target Profit (Mu)	With Target Profit (Mu)	Without Target Profit (Mu)
Buyer's profit	35.00	60.00	75.26	70.38	85.38
Manufacturer's profit	25.00	80.00	65.09	90.38	75.38

Total channel profit with minimum credit to the buyer CP_{min} = 140.35 Mu

Total channel profit with maximum credit to the buyer CP_{max} = 160.76 Mu

Profit distribution between the channel members is given in Table 3.1

Though quantity discount/credit contract has been suggested as a coordinating mechanism in the literature, the authors have not dealt with the risks arising from the overstock that is a major problem commonly observed in many supply chains, particularly when demand is uncertain. Therefore, another form of contracts known as return policies/buyback (BB) contract has been designed by researchers to manage retail overstock. In the next section, a BB contract is discussed.

3.3.4 Buyback Contract

A BB contract (also known as return policy) is another important contract discussed in the literature and practiced by the business organization to coordinate a responsive SC. Return policies allow the retailer to return a certain amount of unsold goods to the manufacturer at the end of the selling season for a partial rebate. Returns policies are common in the distribution of commodities with uncertain demand, such as books, magazines, newspapers, recorded music, computer hardware and software, greeting cards, and pharmaceuticals (Padmanabhan and Png, 1995). This type of contract is commonly suggested to mitigate the risk of retailer's overstock that normally arises due to the uncertain nature of demand, faulty planning and forecasting, and poor purchasing practices. Because of the above-mentioned reasons, a retailer may order less than the system optimal quantity. A BB contract allocates the risk between the supplier and the retailer, and the retailer can return the goods to the supplier and get some money back.

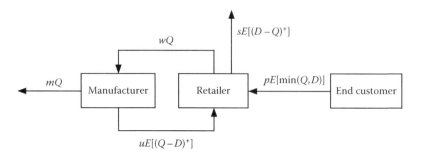

FIGURE 3.3
Flow of payments in a simple supply chain under buyback (BB) contract.

In this contract, the supplier charges a little more for the wholesale price and in return guarantees to give back in the form of a per unit buyback price to the retailer. The buyback price should be less than the wholesale price to make the contract feasible. There exist five types of return policies: full returns for full credit, full returns for partial credit, partial returns for full credit, partial returns for partial credit, and no returns (Wang et al., 2007). The most commonly used return policy is full return for partial credit, which is widely acknowledged as the BB contract.

We have shown in Figure 3.3 the flow of payment when the manufacturer considers a BB contract. As returns are permitted for a certain buyback price at the end of the selling season, the retailer obtains an amount $uE\left[(Q-D)^{+}\right]$ for the leftover items and is shown as inflow to the retailer. The notation used here is similar to that used in a price-only contract. The return price u is smaller than the wholesale price p. The profit equations for retailer and manufacturer under a BB contract can be written as follows. (The notation W_{BB} refers to the wholesale price under a buyback contract.)

$$\Pi_{R_{BB}} = pE\left[\min\left(Q,D\right)\right] - w_{BB}Q + uE\left[(Q-D)^{+}\right] - sE\left[(D-Q)^{+}\right]$$

$$\Pi_{M_{BB}} = \left(w_{BB} - m\right)Q - uE\left[(Q-D)^{+}\right]$$

This contract has been studied from both fixed price newsvendor setting and price-setting newsvendor setting. Pasternack (1985) is one of the front-runners of the BB contract who have applied for a short life-cycle product that has a fixed retail price but stochastic retail demands. He has shown that under the fixed price newsvendor framework, both no returns and full returns for full credit are suboptimal for SC coordination, while full returns for partial credit can achieve SC coordination. This work was extended by many authors relaxing the assumptions and incorporating new constraints. Kandel (1996) extended the work of Pasternack (1985) by introducing the price-sensitive stochastic demand and showed that the SC cannot be coordinated by return policy without retail price maintenance (i.e., allowing the manufacturer to dictate the retail price).

A returns policy from the manufacturer's perspective under retail competition was studied by Padmanabhan and Png (1995). They showed that a returns policy intensifies retail competition and reduces retailer margins. The authors considered two types of return policies in their analysis: no returns and full returns for full credit. Using the price-sensitive demand function, Padmanabhan and Png (1997) further studied the strategic effect of return policy on retailers' competition and highlighted its profitability implications for manufacturers. They concluded that the wholesale price when returns are acceptable should include an insurance premium. Prices would therefore be higher than those when returns are not honored. Emmons and Gilbert (1998) investigated the role of returns policies in the setting that the retailer has control over both the retail price and inventory decisions. They considered a price-sensitive multiplicative model of demand uncertainty for catalog goods and demonstrated that uncertainty tends to increase the retail price. They showed that there exists a range of wholesale prices such that offering a positive buyback price may benefit both the supplier and the retailer. However, their analysis is restricted to a uniform demand distribution.

A risk-free returns policy taking the viewpoint of the manufacturer, who sells a short life-cycle product to a single retailer, was studied by Webster and Weng (2000) and established the conditions under which such a policy exists. They pointed out that when compared to no returns, the retailer's expected profit is increased and the manufacturer's profit is at least as large as when no returns are allowed. Donohue (2000) studied returns policies in a SC model with multiple production opportunities and improving demand forecasts. Taylor (2002) investigated a SC combining both buyback and target sales rebate contracts when demand is sensitive to a retailer's sales effort. He introduced channel rebate as the coordination mechanism for the SC and considered two forms of rebate: linear rebate and target rebate. In the linear rebate case, a rebate is paid for each unit sold and a target rebate is paid for each unit sold beyond a specified target level. The author has shown that when demand is not influenced by sales effort, a properly designed target rebate achieves channel coordination. The author distinguished rebate from a reduction in manufacturer's wholesale price by mentioning that reduction in price caused by the rebate is only realized if the item is sold to an end user; therefore, it has direct impact on retailer's sales effort.

Following the same framework of Padmanabhan and Png (1995), Tsay (2002) analyzed how risk sensitivity affects the use of return policies in a SC. He found that a risk-sensitive retailer does not always prefer the right to return excess product for full credit, and a risk-sensitive manufacturer does not always oppose this, when the wholesale price is fixed. Su and Shi (2002) proposed a QD model with returns contract from the perspective of the manufacturer, in which a menu of discount–return combinations is proposed upon which the manufacturer can make inventory decisions. Both these studies provided frameworks that specifically addressed the returns contract from the manufacturer's perspective, and both assumed that the manufacturer would accept returns only when the profit after returns was no different from the no-returns scenario. However, they ignored the fact that the manufacturer might have no incentive to accept returns because the profit after returns is not increased despite their need to take a demand risk.

A BB contract alone cannot coordinate the channel as BB reduces the incentive for a retailer's promotional efforts (Krishnan et al., 2004). They suggested a combined BB contract and promotional cost-sharing agreement for coordinating a SC in the event of an observable retailer's effort and associated cost.

The effectiveness of return policies in the price-setting newsvendor framework was studied by Granot and Yin (2005). Again return policies from the supplier's perspective were studied by Wang et al. (2007) and extended the work of Emmons and Gilbert (1998) by considering a broader family of demand distributions. They showed the impact on the returns policy by changing the demand patterns in certain ways under the assumption of the exogenous retail price. Wang and Webster (2007) considered a decentralized SC in which a single risk-neutral manufacturer is selling a perishable product to a single loss-averse retailer. They investigated a returns policy with a gain/loss sharing provision to coordinate the SC.

Choi et al. (2008) studied the mean-variance analysis of SC under a return policy. They illustrated how a return policy can be applied for managing the SC to address issues such as channel coordination and risk control. Su and Zhang (2008) studied a decentralized SC with strategic customers who anticipate the seller's future clearance sales at a fixed salvage price and choose the best purchasing time to maximize their expected surplus. They have shown how contractual arrangements can be used to improve the SC performance. Parthasarthy (2011) in his research work has discussed the various models under the BB contract.

3.3.4.1 Some Insights and Limitations

- Though the return policy or BB contract has been suggested as a coordinating mechanism that mitigates the risks of retail overstock, it is impractical in certain situations. For example, the retailer may not have enough cash to procure the items at one go. Hence, the retailer may not be able to purchase the system optimal level stock as the retailer is constrained by cash. Further, the BB contract cannot be cost effective when physically returning products is costly or when SC members have different salvage values for unsold products (Tsay, 2001).
- Though a BB contract motivates the manufacturer to provide incentives for the retailer to stock more, it may not be practical in certain situations. For example, in the video rental industry, it is not practical

for the video rental stores to return excess inventory of old DVDs to the manufacturer (distributor). This may have triggered the idea for the manufacturer to develop a risk-sharing scheme in the form of a revenue-sharing (RS) contract. The RS contract can be specified by the wholesale price and the RS fraction (i.e., the fraction of revenue retained by the retailer).

3.3.5 Revenue-Sharing Contract

A RS contract has been a great success in the video cassette rental industry. It has gained popularity owing to higher product availability. Cachon and Lariviere (2005) have demonstrated how the newly released videos are successfully handled by Blockbuster Inc., a large video retailer, employing RS contract. Earlier, when operating under a conventional agreement, Blockbuster consistently received a lot of customer complaints for the poor availability of newly released videos. It came back with the successful RS formula and increased its market share of video rentals from 24% in 1997 to 40% in 2002. Mortimer (2000) estimated that the RS contract increased the video rental industry's total profit by 7%.

Under a typical RS contract, a supplier charges a retailer a wholesale price per unit that is usually less than the per unit cost of production, but in return receives a fraction of revenue generated from sales at a retail level (Cachon and Lariviere, 2005). In the RS contract, the retailer's and the supply chain's objectives are perfectly aligned. Because the total SC profits are higher compared to the traditional wholesale price contract, partners can choose RS fraction (α) such that both parties are benefitted. The choice of α depends on several factors, including the bargaining power of the supplier and the retailer. The flow of payment in RS contract is shown in Figure 3.4 and detailed discussion can be found in Sarathy (2010). Similar to earlier cases of BB contract, the retailer generates an amount $pE\big[\min(Q,D)\big]$ due to the proceeds of the sale. When RS contract is in place, the retailer retains $rpE\big[\min(Q,D)\big]$ and remits the balance $(1-r)pE\big[\min(Q,D)\big]$ to the manufacturer as per the contract.

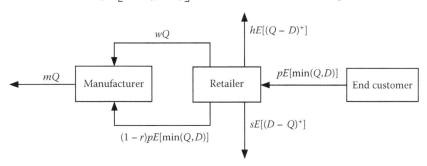

FIGURE 3.4
Flow of payments in a supply chain under revenue-sharing contract.

The retailer incurs a penalty of $hE\left[(Q-D)^+\right]$ and $sE\left[(D-Q)^+\right]$ in the case of overage and shortage, respectively.

Under RS contract, for the order quantity, the profit functions are given by the following equations. The notation W_{RS} refers to the wholesale price under a RS contract:

$$\Pi_{R_{RS}} = rpE\left[\min(Q,D)\right] - w_{RS}Q - hE\left[(Q-D)^+\right] - sE\left[(D-Q)^+\right]$$

$$\Pi_{M_{RS}} = (1-r)pE\left[\min(Q,D)\right] + (w_{RS}-m)Q$$

Many authors have studied RS as an incentive mechanism to coordinate a SC. Pasternack (2001) has investigated the situation in which a vendor, under the newsvendor framework, has the option of purchasing the item outright or obtaining the item through a RS agreement with the manufacturer. The specific interests of this study are the implications of the vendor having limited funds available for purchasing the item. The RS contract in the context of a perfectly competitive market faced by the downstream retailers is studied by Dana and Spier (2001). They have demonstrated that the RS contract can induce downstream firms to choose channel optimal actions. Other related works include Mortimer (2000) and Gerchak et al. (2001).

Wang and Gerchak (2003) and Gerchak and Wang (2004) have considered a RS contract to coordinate assembly systems. Gerchak and Wang (2004) have pointed out that RS alone cannot coordinate the decentralized assembly system including an assembler/retailer and its suppliers. To achieve coordination, they have proposed a revenue plus surplus subsidy scheme for an assembly system where, in addition to a share of revenue, a supplier is partially paid by the assembler for its delivered components that are not sold. Surplus subsidies transfer some of the risk of demand uncertainty from the suppliers to the assembly unit. Three stages of SC coordination through RS contract were studied by Giannoccaro and Pontrandolfo (2004). This model allows the system efficiency to be achieved as well, as it could improve the profits of all SC members by tuning their contract parameters.

Supply chain performance using consignment contract with RS in the context of a single supplier and retailer was studied by Wang et al. (2004). Under the contract, the retailer provides the manufacturer with a marketing medium such as the physical shelf-space in a retail store or the Internet marketplace of Amazon.com, and so forth, for selling their products while the manufacturers retain the ownership of the goods and decide delivery quantities and retail prices for their products. For each item sold, the retailer deducts an agreed percentage from the selling price and remits the balance to the supplier. They have shown that under such a contract, both the overall performance and the performance of individual firms depend critically on price elasticity of demand and the retailer's share of channel cost. They have also pointed out that a consignment arrangement with RS naturally favors the retailer. Because no payment to the supplier is made until the item is sold, the retailer has no money tied up in inventory and bears no risk associated with demand uncertainty.

SC coordination under competition is another new direction of research. Cachon and Lariviere (2005) have studied the RS contract in a general SC model with revenues determined by each retailer's purchase quantity and price. They have demonstrated that RS coordinates a SC with a single retailer (where the retailer chooses optimal price and quantity) and arbitrarily allocates the supply chains' profit. They have compared RS contract to number of other SC contracts and found that RS contract is equivalent to BB contract in the fixed price newsvendor setting and equivalent to price discounts in the price-setting newsvendor setting. They have also found that a RS contract coordinates the SC with retailers competing in quantities (e.g., Cournot competitors or competing newsvendors with fixed price). Despite numerous merits, they have also identified several limitations of RS contract. First, they have pointed out that while the RS contract coordinates retailers that compete on quantity, it does not coordinate retailers that compete on both price and quantity. Second, they argue that the gains from RS contract over a simpler PO contract may not cover revenue sharing's additional administrative expenses. In particular, revenue sharing's incremental improvement over wholesale price contract diminishes as the revenue function becomes more concave or as retail competition intensifies. Third, a RS contract fails to coordinate the supply chain when the demand is influenced by a retailer's effort that is noncontractable and costly (refer to Parthasarahy, 2011).

Tang and Deo (2005) have determined under the RS contract the conditions for wholesale price and RS fraction under which the retailer will obtain a higher profit. Yao et al. (2008a) have found that the provision of RS in a contract can obtain better performance than a PO contract. Yao et al. (2008b) have found that a RS contract can improve SC performance. However, the benefits of RS contract for different SC partners vary because of the impact of demand variability and price-sensitivity factors.

Ha and Tong (2008) have studied a RS contract assuming the demand to be sensitive to service. They have incorporated demand-enhancing service into their model. They have considered a supplier–customer relationship where the customer faces a typical newsvendor problem of determining perishable capacity to meet uncertain demand. Wang et al. (2008) have studied RS contract in a SC with fuzzy demand and proved the effectiveness of a RS contract. Qin and Yang (2008) have analyzed a RS contract using a two-stage Stackelberg game and revealed that for the SC to be more profitable, the member who keeps more than half the revenue should serve as the leader of the Stackelberg game.

The issue of channel coordination for a SC facing stochastic demand, which is sensitive to both sales effort and retail price, was investigated by He et al. (2009). In the standard newsvendor setting, the returns policy and the RS contract have been shown to be able to align incentives of the supply chain's members so that the DSC behaves as the integrated one. When the demand is influenced by both retail price and retailer sales effort, none of the above traditional contracts can coordinate the SC. To resolve this issue, authors have explored a variety of other contract types including joint return policy with RS contract, return policy with sales rebate and penalty (SRP) contract, and RS contract with SRP. They found that only the properly designed returns policy with SRP contract is able to achieve channel coordination and lead to a Pareto improving the win–win situation for SC members. Linh and Hong (2009) have studied channel coordination through a RS contract between a single retailer and a single supplier in a two-period newsboy setting. They found that wholesale prices are set lower than retail prices, and the optimal RS ratio is linearly increasing for the retailer in wholesale prices. More recently, Pan et al. (2010) investigated RS and PO contracts under different channel power structures and analyzed the outcomes under different scenarios to discover whether it is beneficial to use a RS contract.

3.3.5.1 Some Insights and Limitations

- In RS contracts, the wholesale price does not depend on the selling price. But in a BB contract, the wholesale price and the return price depend on the selling price. Hence, in a BB contract the supplier should set wholesale and BB price per unit as functions of the final selling price.

- A BB contract is flexible in terms of how the profits are shared between the supplier and the retailer. For every RS contract, there is an equivalent BB contract, and vice versa.

- A RS contract creates additional cost and administrative burden compared to the straightforward wholesale price contract. This takes an organizational effort to set up the deal and follow its progress. It is worthwhile to go into such contracts only if the increase in profits is relatively large compared to the additional cost and the administrative effort.

- In case of multiple retailers, coordination is not guaranteed unless the supplier has the flexibility to offer different contracts to different retailers. Unfortunately, such discriminations may not always be possible due to legal considerations.

- A RS contract loses its appeal if the revenues depend on the retailer's sales effort. For a retailer who is taking only a fraction of the revenues he or she generates, the incentive to improve sales goes down.

3.3.6 Quantity Flexibility Contract

In quantity flexibility (QF) contract, the supplier provides a full refund for returned (unsold) items as long as the number of returns is no larger than a certain quantity. Thus, this contract gives full refund for a portion of the returned items, whereas a BB contract provides partial refund for all returned items. Further, the QF contract allows the buyer to obtain different quantities than the previous estimate (Lariviere, 1999). A QF contract can be established in a different form such as a minimum purchase quantity contract, backup agreement that allows a buyer to purchase a higher but more limited quantity than his or her initial order quantity (Epen and Iyer, 1997) or special contract that establishes terms, where the buyer needs to purchase a minimum quantity and the supplier needs to deliver up to a certain quantity when the demand exceeds the forecast (Tsay, 1999).

3.3.6.1 Contract with Advanced Commitment

In a most simplistic inventory model, it is seen that a buyer can place an order at any time for any amount at a fixed price or at some discounts. It gives the buyer a great deal of flexibility, but on the other hand, it increases the supplier's uncertainty. So, there is a need to strike a balance between the flexibility of the buyer and increase in uncertainty of the supplier. With this argument, some authors have studied the advanced commitment of order from the buyer for which the supplier gives incentive in the form of discounts, refund facility, or credit term to reduce the uncertainty of his or her demand. Tsay (1999) has modeled a decentralized SC where quantity flexibility is coupled with the customer's commitment to purchase no less than a certain percentage and at the same time, the supplier's guarantee to deliver up to a certain percentage. In this way, author has given weightage to protect the interest of the members.

Tomlin (2000), on the other hand, in his well-researched work has mentioned that though QD are prevalent both in practice and in the operations management literature, yet correctly priced quantity premium to the supplier enables the manufacturer to capture the total SC profit. In quantity premium contract, the manufacturer pays a higher average unit cost as the order size increases. Such a contract may seem counterintuitive as the buyer often expects to be rewarded for placing larger orders. Shi and Su (2002) have considered a two-stage SC where demand is stochastic and the buyer can return the unsold product to the supplier. Considering a rational supplier, who otherwise is willing to produce to maximize his or her profit only, the authors have shown that a retailer should offer an option premium to induce the manufacturer to produce larger quantity. This reduces the inefficiency caused by decentralized control.

3.3.6.2 Some Insights and Limitations

- An interesting insight gathered from here is that by providing correctly priced quantity premium by manufacturer to supplier enables the manufacturer to capture total SC profit (Tomlin, 2000).

- Another important issue of concern in this category of literature is that when the buyer has made an advance commitment for his or her order, then what is the guarantee that the supplier will be able to provide the required quantity in time? This critical issue needs further attention.
- Except for a few, many of the models developed with advanced commitment from the buyer do not take into consideration the updating of the buyer's demand.

3.3.7 Take-It-or-Leave-It Contract

In this contract, the supplier compels the retailer to purchase centralized optimal order quantity at the optimal wholesale price of the supplier. Here, the supplier captures 100% of the CSC profits. The supplier offers the contract and grabs the entire CSC profit as the first-mover advantage. This contract requires a very powerful supplier.

3.3.8 Marginal Pricing Contract

In this contract, the supplier sets a wholesale price equal to its marginal price, so that the retailer grabs the entire CSC profit. This contract requires a very powerful retailer.

It is important to note here that a take-it-or-leave-it contract would require a very powerful supplier, whereas the marginal pricing contract would require a very powerful retailer. These are the two extreme cases, and in reality, neither the supplier nor the retailer is so powerful in general to dictate such contract terms. These two contracts are the extreme cases of SC contracting and are not easily implementable when both supplier and retailer have similar bargaining powers.

3.3.9 Two-Part Tariff

In a two-part tariff contract, a supplier offers the buyer a constant unit wholesale price and a fixed fee, where the buyer chooses order quantity based on internal cost structure, the wholesale price, and the fixed fee offered in the contract. The fixed fee determines the allocation of profits between the supplier and the retailer. Because any allocation is possible, the contract is flexible.

3.4 Coordination under Asymmetric Information

Despite recent advances in information technology and the trend toward sharing information among SC partners, information asymmetry remains an important feature of most of the real SC partnerships. In practice, many business organizations do not like to share business-related information with other members, and this information asymmetry is of two types: demand information and cost-related information. In addition to that, another important question is how much information one should share with his or her SC partner without taking any risk of potential exploitation. This is one of the key issues in SC coordination. Credibility is another key issue in exchange of information between the SC partners. Therefore, when the parties exchange information, it is a key question as to whether a receiver of information should believe in the honesty of the reported information. In particular, it is true when the informed party has an incentive to distort his or her information to influence the receiver's action (Cachon and Lariviere, 2001). They have taken this point into consideration in their study and have proposed contracts that promote sharing of information credibly, particularly demand forecasts information so that an upstream partner can correctly take his or her capacity expansion decisions when the supplier cannot observe the retailer's inventory position.

Few authors have investigated the SC coordination problem when members are less informed about each other or when there is incomplete information between the members of the SC. Desiraju and Moorthy (1997) have examined how the coordination dynamics could be affected by the information asymmetry between the manufacturer and the supplier. Corbett and Tang (1999) in their work have investigated various contracts under single-buyer and single-supplier settings with the presence of asymmetric information. The focus of their study lies on supplier-initiated contract and value to the supplier of getting better information about the buyer's cost structure. They investigated the impact of different types of contracts and information asymmetry on supplier's and buyer's profits, respectively.

Corbette and Groote (2000) in their study on channel coordination have considered the information asymmetry factor in their models. They have derived the supplier's optimal quantity discount scheme when the buyer holds private information about his or her costs structure. Here, the authors have compared the supplier's optimal contract under full information and under asymmetric information. Corbett (2001) demonstrated in this study that when both incentive conflict and asymmetric information are present in a supply chain, then the overall performance of the supply chain is reduced. Ha (2001) in his paper has considered the problem of designing a contract to maximize the supplier's profit in a one-supplier one-buyer supply chain for a short life-cycle product where the supplier is having private information about his or her manufacturing cost and the customer is facing stochastic demand. The author has determined the customer optimal ordering policy and has shown that under asymmetric information, it is no longer optimal for the customer to induce the system-wide optimal solution. Assumptions made in the model are that demand for the finished product is stochastic and price sensitive and only its probability distribution is known. Further, he took two simplistic assumptions in his model, namely a bilateral monopoly relationship with one supplier and one buyer and a newsvendor setup. When the supplier has complete information on the marginal cost of the buyer, the author has shown that several simple contracts can induce the buyer to choose order quantities that attain the single firm profit-maximizing solution resulting in the maximum possible profit for the supplier.

Sucky (2006) has addressed the two-stage SC coordination problem under asymmetric information with an assumption that the buyers could have only two sets of predefined cost structures (all sets are deterministic) and the seller does not know in which particular set a particular buyer belongs. The author has developed a pricing scheme to accommodate the buyers in any of the two predefined clusters. However, the limitation of the model is that it is not flexible enough to address a wide range of cost structures that a buyer can possibly have.

Burnetas et al. (2007) have examined QDs in single-period supply contracts with asymmetric demand information. In contrast to much of the work that has been done on single-period supply contracts, the authors have assumed that there are no interactions between the supplier and the buyer after demand information is revealed and that the buyer has better information about the distribution of demand than does the supplier. They characterized the structure of the optimal discount schedule for both all-unit and incremental discounts and showed that the supplier can earn larger profits with an all-unit discount. They have also suggested that the QD contract can be used in conjunction with other mechanisms, such as returns policies, quantity flexibility, and price protection that can offer potential scope for future research.

Sarmah et al. (2008b) have developed a coordination model considering credit option as a coordination mechanism when the firms involved in the chain do not have complete information about each other's cost structure and target profit. They have shown in such a situation how the two firms can coordinate despite having limited information. Further, in a decentralized SC where the firm has its individual target profit, how can the surplus generated in the system be distributed satisfying each firm's target profit? The authors have developed a negotiation mechanism to share the surplus profit generated due to coordination.

Another class of literature deals with uncertain information by applying fuzzy set theory to estimate the uncertain variables, particularly in a case where there are no adequate historical data available to derive the stochastic distribution for a particular parameter. Application of fuzzy sets deals with possibility distribution unlike probability distribution in stochastic processes. Mahata et al. (2005) have investigated the JELS model for both purchaser and vendor in fuzzy sense. The authors have extended Banerjee's (1986) model with the assumption that the order quantity for the purchaser is a fuzzy variable while the rest of the parameters are deterministic. However, the model does not include a quantity discount policy to coordinate between the buyer and the vendor. Moreover, the authors have not derived any mechanism to share the coordination benefits among the different partners.

Sinha and Sarmah (2008) have developed a coordination mechanism under asymmetric information that allows a system to perform as closely as that of centralized SC under complete information. The authors have used fuzzy set theory to estimate the uncertainty associated with different input parameters including the holding cost information of the buyer.

3.4.1 Some Insights and Limitations

- When information asymmetry and incentive conflict are present in a SC then it reduces the overall performance of the SC.

- How the conflict between the two parties of the SC will be resolved through negotiation is an important issue of concern, particularly when both parties have limited information about each other.

- Benefits of information sharing depend on the structure of the particular SC and method of its working. Benefits depend on various parameters of SC such as capacity and variability in demand.

- Another important issue of concern is the credibility of the shared information between the different parties of the supply chain.

3.5 Conclusions and Directions for Future Research

Over a period of time, the relationships between different business entities have undergone significant changes with increasing emphasis on coordination and information sharing. People have realized the benefits of SC coordination, and the advantages are enormous. It is now well established that coordination and collaboration help business organizations in reducing inventory, increasing sales, lowering costs, increasing revenue, improving forecast accuracy, reducing lead time, ensuring timely delivery, and offering better customer service.

In this chapter, we discussed various mathematical models developed to coordinate a SC under various situations. Some of the future research directions related to SC coordination modeling are as follows:

- In the current business environment, development of contractual agreements between the members of the SC is an important issue. Often the nature of a contractual agreement affects SC coordination. The agreement should motivate the members to implement an integrated decision to improve the efficiency of the SC. There is no universal coordination mechanism that can be applied for all kinds of SC. Coordination mechanisms are situation dependent. A coordination mechanism that performs very well in one kind of situation may not necessarily perform equally well in another situation. Therefore, different mechanisms are evolving over time to handle different situations. Even in certain kinds of business situations, one mechainsm alone is not adequate to coordinate a SC; therefore, hybrid coordination mechanisms are developed. Development of a hybrid coordination mechanism is an interesting future research direction.

- It is well documented in the literature that due to coordination, a surplus is generated in the system. Coordination between two parties is not a zero-sum game where one gains only at the expense of the other. For coordination to be effective, each party may want a certain amount of surplus generated due to coordination and therefore could set a certain target profit. Development of analytical models considering target profit for both the parties to have effective coordination is an interesting research direction for the future.

- Incorporation of a negotiation process to divide the benefits of coordination is also an important future research direction. The negotiation process focuses on dynamic sharing of surplus between the two parties, where both can take part in the decision making and negotiation ends with a win–win situation for both. This is considered to be a superior solution compared to a predetermined static division of surplus through side payment strategy.

- Development of a model with the assumption of complete available information does not reflect the real situation. The comfortable situation of complete information sharing between business entities hardly exists in the real world. Under such circumstances, it is important to develop models that can take care of the coordination problem under asymmetric information. Development of a negotiation process under asymmetric information is an important future research direction of study.

- Another area of SC coordination that has drawn the attention of researchers is development of a suitable mechanism to coordinate the logistic processes that are controlled by various companies. Swenseth and Godfrey (2002) have reported that often about 50% of total annual logistics cost of a product can be attributed to transportation cost. Therefore, for the overall performance improvement of the SC, there is a need to develop a coordination mechanism to coordinate the logistics processes between the various parties. Particularly, in the multibuyers case where buyers are located in different geographical regions, individual shipments to the buyers by the vendor increases the total system cost. In such a situation, coordinated shipment from the vendors to multiple buyers helps to reduce channel cost.

- SC coordination and competition together have not received adequate attention from researchers. Specifically, there may be multiple sellers who compete on selling a similar kind of product in the same market, and for that there may be different SC with different channel structures. From the perspective of economic theory, a large number of research papers are available on market competition which mostly deal with either quantity-competition (Cournot) or price-competition (Bertrand), and their primary focus is on applying game theory to derive equilibrium under varied assumptions. Although a few studies have been carried out from the perspectives of marketing and operations management, considering the different settings of SC, there is a need to understand the impact of different SC settings on market parameters such as demand, product type, and price.

- Today's SCs are vulnerable to different kinds of risks; therefore, risk mitigation in SC is an important issue. An interesting research direction is the development of a coordination mechanism to mitigate the SC risks.

References

Banerjee, A. 1986. A joint economic lot size model for purchaser and vendor. *Decision Science*, 17(3), 292–311.

Banerjee, A., and Burton, J.S. 1994. Coordinated vs. independent inventory replenishment policies for a vendor and multiple buyers. *International Journal of Production Economics*, 35(3), 215–222.

Banerjee, A., Kim, S.L., and Burton, J. 2007. Supply chain coordination through effective multi-stage inventory linkages in a JIT environment. *International Journal of Production Economics*, 108(1–2), 271–280.

Ben-Daya, M., and Al-Nassar, A. 2008. An integrated inventory production system in a three-layer supply chain. *Production Planning and Control*, 19(2), 97–104.

Benton, W.C., and Park, S. 1996. A classification of literature on determining the lot size under quantity discounts. *European Journal of Operational Research*, 92(2), 219–238.

Boyaci, T., and Gallego, G. 2002. Coordinating pricing and inventory replenishment policies for one wholesaler and one or more geographically dispersed retailers. *International Journal of Production Economics*, 77(2), 95–111.

Burnetas, A., Gilbert, S.M., and Smith, C.E. 2007. Quantity discounts in single-period supply contracts with asymmetric demand information. *IIE Transactions*, 39(5), 465–479.

Cachon, G., and Fisher, M. 2000. Supply chain inventory management and the value of shared information. *Management Science,* 46(8), 1032–1048.

Cachon, G.P. 2001. Stock wars: Inventory competition in a two-eschelon supply chain with multiple retailers. *Management Science,* 49(5), 658–674.

Cachon, G.P., and Lariviere, M.A. 2005. Supply chain coordination with revenue-sharing contracts: Strengths and limitations. *Management Science,* 51(1), 30–44.

Cachon, G.P., and Lariviere, A.M. 2001. Contracting to assure supply: How to share demand forecasts in a supply chain. *Management Science,* 47(5), 629–646.

Cachon, G.P., and Zipkin, P. 1999. Competitive and cooperative inventory policies in a two stage supply chain. *Management Science,* 45(7), 936–953.

Cárdenas-Barrón, L.E. 2007. Optimizing inventory decisions in a multi-stage multi-customer supply chain: A note. *Transportation Research Part E: Logistics and Transportation Review,* 43(5), 647–654.

Chapman, C.B., Ward, S.C., Cooper, D.F., and Page, M.J. 1985. Credit policy and inventory control. *Journal of Operational Research Society,* 35(12), 1055–1065.

Chen, F., Federgruen, A., and Zheng, Y. 2001. Coordination mechanisms for a distribution system with one supplier and multiple retailers. *Management Science,* 47(5), 693–708.

Chen, L.H., and Kang, F.S. 2007. Integrated vendor–buyer cooperative inventory models with variant permissible delay in payments. *European Journal of Operational Research,* 183(2), 658–673.

Chiang, W.C., Fitzsimmons, J., Huang, Z., and Li, S. 1994. A game theoretic approach to quantity discount problem. *Decision Science,* 25(1), 153–166.

Choi, T.-M., Li, D., and Yan, H. 2008. Mean-variance analysis of a single supplier and retailer supply chain under a returns policy. *European Journal of Operational Research,* 184(1), 356–376.

Christy, P.D., and Grout, R.J. 1994. Safeguarding supply chain relationships. *International Journal of Production Economics,* 36(3), 233–242.

Chu, W.H.J., and Lee, C.C. 2006. Strategic information sharing in a supply chain. *European Journal of Operational Research,* 174, 1567–1579.

Cooper, M.C., Lambert, D.M., and Pagh, J.D. 1997. Supply chain management: More than a new name for logistics. *International Journal of Logistics Management,* 8(1), 1–13.

Corbett, C. J., and Tang, C.S. 1999. Designing supply contracts: Contract type and information asymmetry. In S. Tayur, M. Magazine, and R. Ganeshan (Eds.), *Quantitative Models for Supply Chain Management.* Heidelberg: Kluwer Academic, 269–297.

Corbette, C.J., and Groote, X. De. 2000. A supplier's optimal quantity discount policy under asymmetric information. *Management Science,* 46(3), 444–450.

Costantino, F., and Di Gravio, G. 2009. Multistage bilateral bargaining model with incomplete information—A fuzzy approach. *International Journal of Production Economics,* 117(2), 235–243.

Dana, Jr., J.D., and Spier, K.E. 2001. Revenue sharing and vertical control in the video rental industry. *Journal of Industrial Economics,* 49(3), 223–245.

Davis, R.A., and Gaither, N.A. 1985. Optimal ordering policies under condition of extended payment privileges. *Management Science,* 31(4), 499–509.

Desiraju, R., and Moorthy, S. 1997. Managing a distribution channel under asymmetric information with performance requirements. *Management Science,* 43(12), 1628–1644.

Donohue, K. 2000. Efficient supply chain contracts for fashion goods with forecast updating and two production modes. *Management Science,* 46(11), 1397–1411.

Ellienhausen, G., Lundquist, E.C., and Staten, M.E. 2003. The impact of credit counseling on subsequent borrower credit usage and payment behavior. Credit Research Center, Georgetown University Working Paper.

Emmons, H., and Gilbert, S.M. 1998. Note. The role of returns policies in pricing and inventory decisions for catalogue goods. *Management Science,* 44(2), 276–283.

Epen, E.G., and Iyer, V.A. 1997. Improved fashion buying with Bayesian updates. *Operations Research,* 45(6), 805–819.

Erhun, F., and Keskinocak, P. 2011. Collaborative supply chain management. In K.G. Kempf, P. Keskinocak, and R. Uzsoy (Eds.), *Planning Production and Inventories in the Extended Enterprise: A State of the Art Handbook.* Heidelberg: Springer, 233–268.

Ertogral, K., and Wu, D.S. 2001. A bargaining game of supply chain contracting. Available at: www.lehigh.edu/ise/documents/01t_002.pdf

Esmaeili, M., Aryanezhad, M.B., and Zeephongsekul, P. 2009. A game theory approach in seller–buyer supply chain. *European Journal of Operational Research,* 195(2), 442–448.

Gerchak, Y., and Wang, Y. 2004. Revenue-sharing vs. wholesale-price contracts in assembly systems with random demand. *Production and Operations Management,* 13(1), 23–33.

Gerchak, Y., Cho, R.K., and Ray, S. 2001. Coordination and dynamic shelf-space management of video movie rentals. Working paper, Department of Industrial Engineering, Tel-Aviv University, Israel.

Giannoccaro, I., and Pontrandolfo, P. 2004. Supply chain coordination by revenue sharing contracts. *International Journal of Production Economics,* 89(2), 131–139.

Goyal, S.K. 1976. An integrated inventory model for a single supplier single customer problem. *International Journal of Production Research,* 15(1), 107–111.

Goyal, S.K. 1985. Economic order quantity under condition of permissible delay in payments. *Journal of Operation Research Society,* 36(4), 335–338.

Goyal, S.K., and Gupta, Y.P. 1989. Integrated inventory model: The buyer vendor coordination. *European Journal of Operational Research,* 41(3), 261–269.

Granot, D., and Yin, S. 2005. On the effectiveness of returns policies in the price-dependent newsvendor model. *Naval Research Logistics,* 52(8), 765–779.

Gurnani, H. 2001. A study of quantity discount pricing models with different ordering structures: Order coordination, order consolidation and multi tier ordering hierarchy. *International Journal of Production Economics,* 72(3), 203–225.

Ha, A. 2001. Supplier buyer contracting: Asymmetric cost information and cut off level policy for buyer participation. *Naval Research Logistics,* 48(1), 41–64.

Ha, A.Y., and Tong, S. 2008. Revenue sharing contracts in a supply chain with uncontractible actions. *Naval Research Logistics,* 55(5), 419–431.

He, Y., Zhao, X., Zhao, L., and He, J. 2009. Coordinating a supply chain with effort and price dependent stochastic demand. *Applied Mathematical Modelling,* 33(6), 2777–2790.

Ingene, A.C., and Parry, E.M. 1995. Coordination and manufacturer profit maximization: The multiple retailer channel. *Journal of Retailing,* 71(2), 129–151.

Issakson, A. 2002. Trade credit in Kenyan manufacturing: Evidence from plant level data. Working paper (www.unido.org./fileadmin/import/userfiles/PuffK/SIN_WPS04.pdf).

Jaber, M.Y., and Osman, I.H. 2006a. Coordinating a two-level supply chain with delay in payments and profit sharing. *Computers and Industrial Engineering*, 50, 385–400.

Jaber, M.Y., Osman, I.H., and Guiffrida, A.L. 2006b. Coordinating a three-level supply chain with price discounts, price dependent demand, and profit sharing. *International Journal of Integrated Supply Management*, 2(1–2), 28–48.

Kandel, E. 1996. The right to return. *Journal of Law and Economics*, 39(1), 329–356.

Khouja, M. 2003. Optimizing inventory decisions in a multistage multicustomer supply chain. *Transportation Research Part E*, 39(3), 193–208.

Kingsman, G.B. 1983. The effect of payment rules on ordering and stockholding in purchasing. *Journal of Operational Research Society*, 34(11), 1085–1098.

Krishnan, H., Kapuscinski, R., and Butz, D.A. 2004. Coordinating contracts for decentralized supply chains with retailer promotional effort. *Management Science*, 50(1), 48–63.

Lal, R., and Staelin, R. 1984. An approach for developing an optimal discount pricing policy. *Management Science*, 30(12), 1524–1539.

Lapide, L. 1998. Are we moving from buyers and sellers to collaborators? *AMR Research Report Supply Chain Management* (July 1).

Lariviere, M.A. 1999. Supply chain contracting and coordination with stochastic demand. In S. Tayur, M. Magazine, and R. Ganeshan (Eds.), *Quantitative Models for Supply Chain Management*. Heidelberg: Kluwer Academic, 233–268.

Lariviere, M.A., and Porteus, E.L. 2001. Selling to the news vendor: An analysis of price only contract. *Manufacturing and Service Operations Management*, 3(4), 293–305.

Lee, J.H., and Moon, I.K. 2006. Coordinated inventory models with compensation policy in a three level supply chain. Lecture Notes in Computer Science, 3rd. Part, LNCS 3982, M. Gavrilova (Ed.), 600–609.

Lee, W. 2005. A joint economic lot-size model for raw material ordering, manufacturing setup, and finished goods delivering. *Omega*, 33(2), 163–174.

Leng, M., and Parlar, M. 2005. Game theoretic applications in supply chain management: A review. *INFORM*, 43(3), 187–220.

Li, S., Huang, Z., and Ashley, A. 1995. Seller–buyer system cooperation in a monopolistic market. *Journal of Operational Research Society*, 46(12), 1456–1470.

Li, S., Huang, Z., and Ashley, A. 1996. Improving buyer–seller system cooperation through inventory control. *International Journal of Production Economics*, 43(1), 37–46.

Linh, C.T., and Hong, Y. 2009. Channel coordination through a revenue sharing contract in a two-period newsboy problem. *European Journal of Operational Research*, 198(3), 822–829.

Lu, Lu. 1995. Theory and methodology: A one vendor multi buyer integrated inventory model. *European Journal of Operational Research*, 81(2), 312–323.

Mahata, G., Goswami, A., and Gupta, D. 2005. A joint economic-lot-size model for purchaser and vendor in fuzzy sense. *Computers and Mathematics with Applications*, 50(10–12), 1767–1790.

Mishra, A.K. 2004. Selective discount for supplier–buyer coordination using common replenishment epochs. *European Journal of Operational Research*, 153(3), 751–756.

Monahan, J.P. 1984. A quantity discount pricing model to increase vendor profits. *Management Science*, 30(6), 720–726.

Mortimer, J.H. 2000. The effects of revenue-sharing contracts on welfare in vertically separated markets: Evidence from the video rental industry. Working paper, University of California at Los Angeles, CA.

Moses, M., and Seshadri, S. 2000. Policy mechanisms for supply chain co-ordination. *IIE Transactions*, 32(3), 245–262.

Munson, L.C., and Rosenblatt, J.M. 1998. Theories and realities of quantity discounts: An exploratory study. *Production and Operations Management*, 7(4), 352–369.

Munson, L.C., and Rosenblatt, J.M. 2001. Coordinating a three level supply chain with quantity discounts. *IIE Transactions*, 33(4), 371–384.

Padmanabhan, V., and Png, I.P.L. 1995. Returns policies: Make money by making good. *Sloan Management Review*, 37(1), 65–72.

Padmanabhan, V., and Png, I.P.L. 1997. Manufacturer's return policies and retail competition. *Marketing Science*, 16(1), 81–94.

Pan, K., Lai, K.K., Leung, S.C.H., and Xiao, D. 2010. Revenue-sharing versus wholesale price mechanisms under different channel power structures. *European Journal of Operational Research*, 203(2), 532–538.

Parthasarathy, G. 2011. Modelling supply chain coordination through contract. Unpublished PhD thesis submitted to the Department of Industrial Engineering and Management, IIT Kharagpur, India.

Pasternack, B.A. 1985. Optimal pricing and return policies for perishable commodities. *Marketing Science*, 4(2), 166–176.

Pasternack, B.A. 2001. The capacitated newsboy problem with revenue sharing. *Journal of Applied Mathematics and Decision Sciences*, 5(1), 21–33.

Pourakbar, M., Farahani, Z.R., and Asgari, N. 2007. A joint economic lot-size model for an integrated supply network using genetic algorithm. *Applied Mathematics and Computation*, 189(1), 583–596.

Qin, Z., and Yang, J. 2008. Analysis of a revenue-sharing contract in supply chain management. *International Journal of Logistics Research and Applications*, 11(1), 17–29.

Sarmah, S.P., Acharya, D., and Goyal, S.K. 2006. Buyer–vendor coordination models in supply chain management: An invited review. *European Journal of Operational Research*, 175, 1–15.

Sarmah, S.P., Acharya, D., and Goyal, S.K. 2007. Coordination and profit sharing between manufacturer and a buyer with target profit under consideration. *European Journal of Operational Research*, 182(3), 1469–1478.

Sarmah, S.P., Acharya, D., and Goyal, S.K. 2008a. Coordination of a single manufacturer/multiple buyer supply chain with credit option. *International Journal of Production Economics*, 111(2), 676–685.

Sarmah, S.P., Acharya, D., and Goyal, S.K. 2008b. Two stage supply chain coordination through credit option in asymmetric information. *International Journal of Logistics Systems and Management*, 4(1), 98–115.

Sheen, G.J., and Tsao, Y.C. 2007. Channel coordination, trade credit and quantity discounts for freight cost. *Transportation Research Part E: Logistics and Transportation Review*, 43(2), 112–128.

Shi, S.C., and Su, T.C. 2002. Improving supply chain efficiency via option premium incentive. *Production Planning and Control*, 13(3), 236–242.

Siajadi, H., Ibrahim, R., and Locher, P. 2006. A single-vendor multiple-buyer inventory model with a multiple-shipment policy. *International Journal of Advanced Manufacturing Technology*, 27, 1030–1037.

Sinha, S., and Sarmah, S.P. 2008. An application of fuzzy set theory for supply chain coordination. *International Journal of Management Science and Engineering Management*, 3(1), 19–32.

Spengler, J. 1950. Vertical integration and anti-trust policy. *Journal of Political Economics*, 58, 347–352.

Stadtler, H. 2009. A framework for collaborative planning and state of the art. *OR Spectrum*, 31(1), 5–30.

Su, C.-T., and Shi, C.-S. 2002. A manufacturer's optimal quantity discount strategy and return policy through game-theoretic approach. *Journal of the Operational Research Society*, 53(8), 922–926.

Su, X., and Zhang, F. 2008. Strategic customer behavior, commitment, and supply chain performance. *Management Science*, 54(10), 1759–1773.

Sucky, E. 2006. A bargaining model with asymmetric information for a single supplier–single buyer problem. *European Journal of Operational Research*, 171(2), 516–535.

Swanson, A. 2003. Ford marks 100th anniversary (http://www.archives.hawaii reporter.com/story.aspx?431cf5aa-eb6c-4433-a358-297b8889781e).

Swenseth, R.S., and Godfrey, M.R. 2002. Incorporating transportation costs into inventory replenishment decisions. *International Journal of Production Economics*, 77(2), 113–130.

Tang, C.S., and Deo, S. 2005. Rental price and rental duration under retail competition. Working paper, UCLA Anderson School.

Taylor, A.T. 2002. Supply chain coordination under channel rebates with sales effort effects. *Management Science*, 48(8), 992–1007.

Thomas, D.J., and Griffin, P.M. 1996. Coordinated supply chain management. *European Journal of Operational Research*, 94(1), 1–15.

Tilson, V. 2008. Monotonicity properties of wholesale price contracts. *Mathematical Social Sciences*, 56(1), 127–143.

Tomlin, B.T. 2000. Supply chain design: Capacity, flexibility and wholesale price strategies. Unpublished PhD thesis submitted to Massachusetts Institute of Technology.

Tsay, A., Nahmias, S., and Agarwal, N. 1999. Modeling supply chain contracts: A review. In S. Tayur, M. Magazine, and R. Ganeshan (Eds.), *Quantitative Models for Supply Chain Management*. Heidelberg: Kluwer Academic, 301–336.

Tsay, A.A. 2001. Managing retail channel overstock: Markdown money and return policies. *Journal of Retailing*, 77(4), 457–492.

Tsay, A.A. 2002. Risk sensitivity in distribution channel partnerships: Implications for manufacturer return policies. *Journal of Retailing*, 78(2), 147–160.

Tsay, A.A. 1999. The quantity flexibility contract and supplier customer incentives. *Management Science*, 45(10), 1339–1358.

Viswanathan, S. 1998. Optimal strategy for the integrated vendor–buyer inventory model. *European Journal of Operational Research*, 105(1), 38–42.

Viswanathan, S. 2001. Coordinating supply chain inventories through common replenishment epoch. *European Journal of Operational Research*, 129(2), 277–286.

Wang, C.X., and Webster, S. 2007. Channel coordination for a supply chain with a risk-neutral manufacturer and a loss-averse retailer. *Decision Sciences*, 38(3), 361–389.

Wang, H., Wang, W., and Kobaccy, K.A.H. 2007. Analysis and design of returns policies from a supplier's perspective. *Journal of the Operational Research Society,* 58(3), 391–401.

Wang, J., Zhao, R., and Tang, W. 2008. Supply chain coordination by revenue-sharing contract with fuzzy demand. *Journal of Intelligent and Fuzzy Systems*: *Applications in Engineering and Technology,* 19(6), 409–420.

Wang, Q., and Wang, R. 2005. Quantity discount pricing policies for heterogeneous retailers with price sensitive demand. *Naval Research Logistics*, 52(7), 645–658.

Wang, Q. 2002. Determination of supplier's optimal quantity discount schedules with heterogenous buyers. *Naval Research Logistics,* 49(1), 46–59.

Wang, Y., and Gerchak, Y. 2003. Capacity games in assembly systems with uncertain demand. *Manufacturing and Service Operations Management*, 5(3), 252–267.

Wang, Y., Jiang, L., and Shen, Z.J. 2004. Channel performance under consignment contract with revenue sharing. *Management Science*, 50(1), 34–47.

Webster, S., and Weng, Z.K. 2000. A risk-free perishable item returns policy. *Manufacturing and Service Operations Management*, 2(1), 100–106.

Wilson, N., Wetherhill, P., and Summers, B. 2000. Business to business credit: A strategic tool for the "New Economy" (http://cmrc.co.uk), Credit Manangement Research Centre.

Woo, Y.Y., Hsu, S.L., and Wu, S. 2001. An integrated inventory model for a single vendor and multiple buyers with ordering cost reduction. *International Journal of Production Economics,* 73(3), 203–215.

Yao, Z., Leung, S.C.H., and Lai, K.K. 2008a. Manufacturer's revenue-sharing contract and retail competition. *European Journal of Operational Research*, 186(2), 637–651.

Yao, Z., Leung, S.C.H., and Lai, K.K. 2008b. The effectiveness of revenue-sharing contract to coordinate the price-setting newsvendor products' supply chain. *Supply Chain Management: An International Journal*, 13(4), 263–271.

Zahir, S., and Sarker, R. 1991. Joint economic ordering policies of multiple wholesalers and a single manufacturer with price dependent demand functions. *Journal of Operational Research Society*, 42, 157–164.

Zhou, Y.W., and Wang, S.D. 2007. Optimal production and shipment models for a single-vendor–single-buyer integrated system. *European Journal of Operational Research*, 180(1), 309–328.

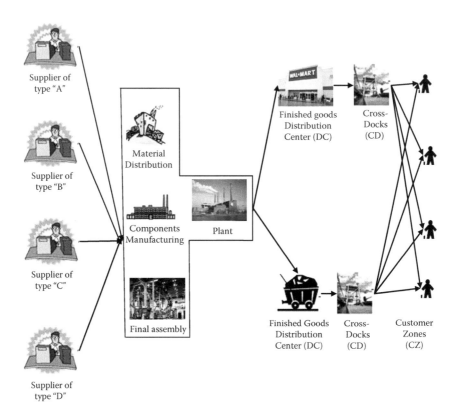

FIGURE 2.1
Integrated network of production, distribution, and logistics activities.

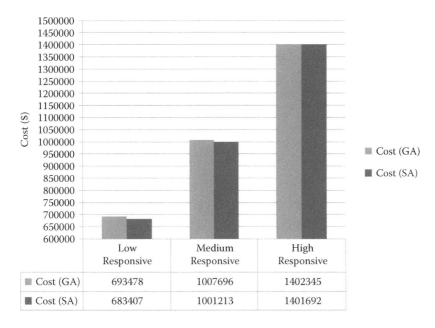

	Low Responsive	Medium Responsive	High Responsive
Cost (GA)	693478	1007696	1402345
Cost (SA)	683407	1001213	1401692

FIGURE 2.4
Comparison of results in terms of cost for problem instance 1.

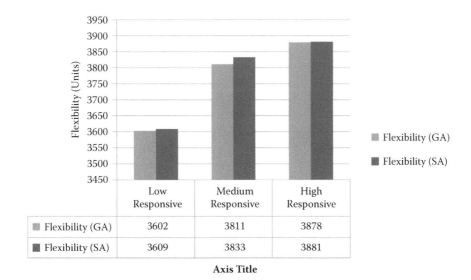

	Low Responsive	Medium Responsive	High Responsive
Flexibility (GA)	3602	3811	3878
Flexibility (SA)	3609	3833	3881

Axis Title

FIGURE 2.5
Comparison of results in terms of flexibility for problem instance 1.

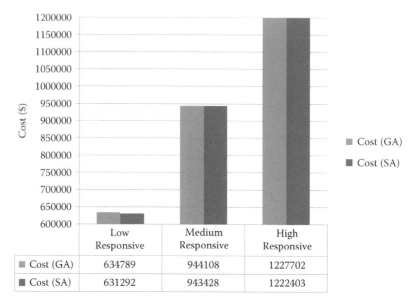

FIGURE 2.6
Comparison of results in terms of cost for problem instance 2.

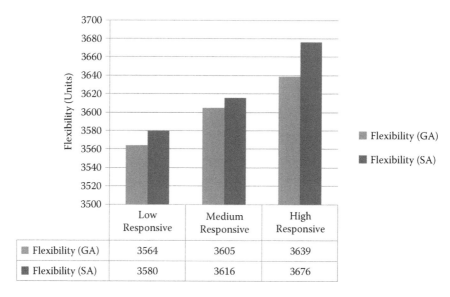

FIGURE 2.7
Comparison of results in terms of flexibility for problem instance 2.

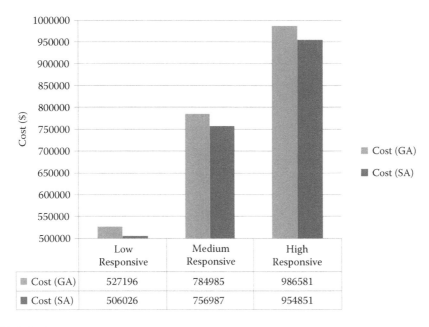

FIGURE 2.8
Comparison of results in terms of cost for problem instance 3.

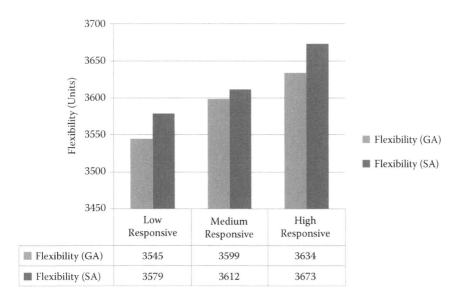

FIGURE 2.9
Comparison of results in terms of flexibility for problem instance 3.

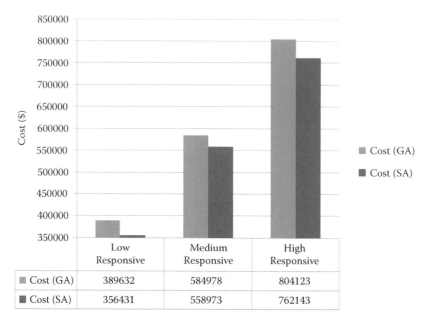

	Low Responsive	Medium Responsive	High Responsive
Cost (GA)	389632	584978	804123
Cost (SA)	356431	558973	762143

FIGURE 2.10
Comparison of results in terms of flexibility for problem instance 4.

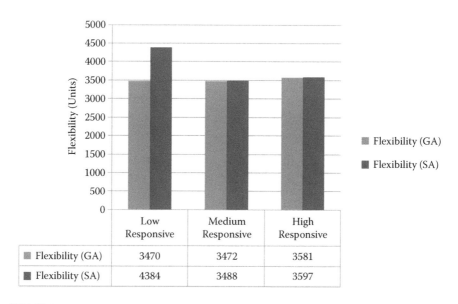

	Low Responsive	Medium Responsive	High Responsive
Flexibility (GA)	3470	3472	3581
Flexibility (SA)	4384	3488	3597

FIGURE 2.11
Comparison of results in terms of flexibility for problem instance 4.

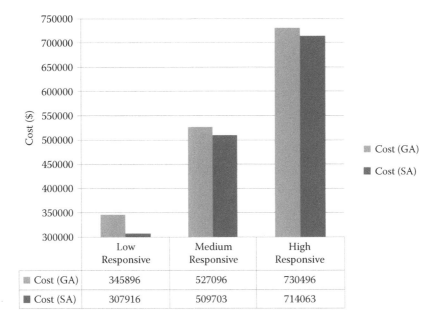

	Low Responsive	Medium Responsive	High Responsive
Cost (GA)	345896	527096	730496
Cost (SA)	307916	509703	714063

FIGURE 2.12
Comparison of results in terms of cost for problem instance 5.

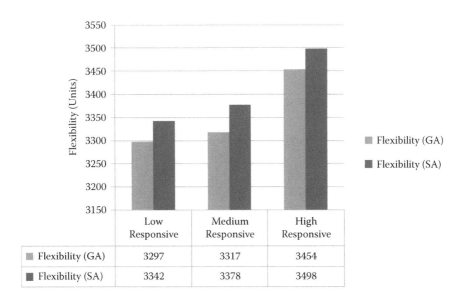

	Low Responsive	Medium Responsive	High Responsive
Flexibility (GA)	3297	3317	3454
Flexibility (SA)	3342	3378	3498

FIGURE 2.13
Comparison of results in terms of flexibility for problem instance 5.

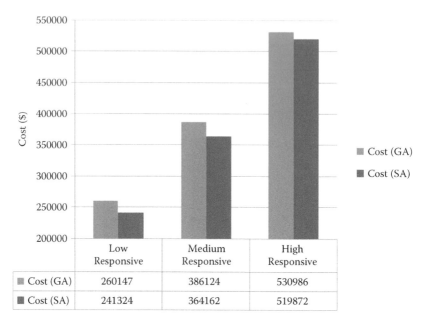

FIGURE 2.14
Comparison of results in terms of cost for problem instance 6.

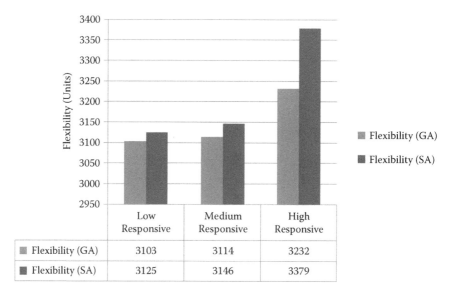

FIGURE 2.15
Comparison of results in terms of flexibility for problem instance 6.

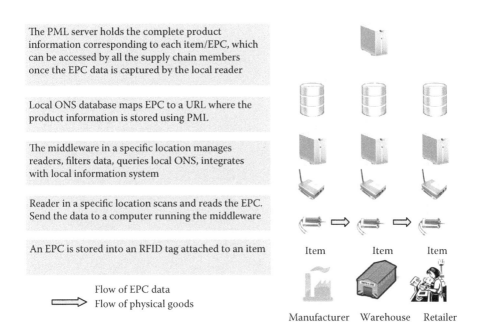

The PML server holds the complete product information corresponding to each item/EPC, which can be accessed by all the supply chain members once the EPC data is captured by the local reader

Local ONS database maps EPC to a URL where the product information is stored using PML

The middleware in a specific location manages readers, filters data, queries local ONS, integrates with local information system

Reader in a specific location scans and reads the EPC. Send the data to a computer running the middleware

An EPC is stored into an RFID tag attached to an item

Item   Item   Item

Flow of EPC data
Flow of physical goods

Manufacturer   Warehouse   Retailer

FIGURE 5.10
Using Electronic Product Code (EPC) infrastructure in the supply chain.

4

System Dynamics Applications in Supply Chain Modeling

4.1 Introduction

Organizations in the twenty-first century have shown increasing interest in efficient supply chain management to achieve competitive advantage in the global marketplace. This is due to the rising cost of manufacturing and transportation, the globalization of market economies, and the ever-increasing demand for diverse products of short life cycles. A supply chain is typically characterized by a forward flow of materials and a backward flow of information. Efficient supply chain management can lead to lower production cost, inventory cost, and transportation cost and improved customer service throughout all stages involved in the chain (Sarimveis et al., 2008).

Different types of supply chain strategies (SCSs) have received increasing attention from both researchers and practitioners. Lee (2002) put forth a demand and supply uncertainty framework that produces four types of SCSs: efficient, risk-hedging, responsive, and agile. Apart from them, there could be many other strategies that could be employed to make a supply chain efficient. It could be too expensive or even impossible to observe the efficacy of these strategies in real supply chains, as some of them may as well turn out to be counterproductive. Supply chain simulation is of use in these situations.

Various alternative methods have been proposed for modeling supply chains. Beamon (1998) grouped these methods into four categories: deterministic models, stochastic models, economic game-theoretic models, and models based on simulation. The deterministic models are utilized when all the parameters are known, while stochastic models are used when at least one parameter is unknown but which follows a probability distribution. The game-theoretic models and models based on simulation involve the evaluation of the performance of various supply chain strategies. The majority of these models are steady-state models based on average performance or steady-state conditions. Static models are insufficient in dealing with the dynamic characteristics of the supply chain system that are due to demand fluctuations, lead-time delays, sales forecasting, and so forth. In particular, they are not able to describe, analyze, and find remedies for a major problem in supply chains, which recently became known as "the bullwhip effect" (Sarimveis et al., 2008).

Supply chain simulation requires the development of a model that suitably represents the problem situation. Performance of the model is then studied to make inferences about the real supply chain. The strength of the simulation approach lies in its ability to deal with complexity of the real supply chains, which may be too difficult to solve by employing analytical techniques. Additionally, simulation approaches are applicable in situations involving uncertainty. Campuzano and Mula (2011) described four approaches to supply chain simulations: spreadsheet-based, system dynamics, discrete-event systems simulation, and business games. While all the other approaches are quite popular, we confine our discussions to the system dynamics approach of supply chain simulations.

One possible use of system dynamics in the supply chain management area is in the modeling of the bullwhip effect—a curse that often hinders the effective performance of a supply chain. The bullwhip effect may be interpreted as an outcome of the strategic interactions among rational supply chain members who represent a series of companies, each ordering goods to its immediate upstream member. In this setting, inbound orders from a downstream member serve as a valuable informational input to upstream production and inventory decisions. The retailer's orders do not coincide with the actual retail sales. Orders to the supplier tend to have larger variance than sales to the buyer, and the distortion propagates upstream in an amplified form, which is referred to as the *bullwhip effect*.

The rest of this chapter is organized as follows. Section 4.2 presents the characteristics of system dynamics. Section 4.3 presents supply chain modeling using system dynamics, while Section 4.4 presents the bullwhip effect and its modeling using system dynamics. Section 4.5 presents a case study on the retailer's inventory. The conclusions are given thereafter.

4.2 Characteristics of System Dynamics

System dynamics is a methodology of system inquiry (Forrester, 1961; Wolstenholme and Coyle, 1983). It helps in carrying out policy experimentation in a continuous time simulation environment with the help of a causal model of a system. System dynamics is a computer-aided approach for analyzing and solving complex problems with a focus on policy analysis and design (Angerhofer and Angelides, 2000). Forrester (1961) stated that system dynamics is a theory of structure and behavior of systems that helps in analyzing and representing, graphically and mathematically, the interactions governing the dynamic behavior of complex socioeconomic systems. System dynamics computer simulation compensates for the deficiencies in mental models. This has been emphasized by psychologists, who have shown that only a few factors contribute while making decisions (Hogarth, 1980). System dynamics cross-fertilizes elements of traditional management, feedback control theory, and computer simulation (Mohapatra et al., 1994). Literature is also available on techniques on modeling large societal systems and the advantage of using system dynamics as a tool for policy experimentation. Forrester (1961), Coyle (1977), Mohapatra et al. (1994), and Sterman (2000) presented good examples in this direction.

System dynamics (SD) is an approach to understanding the behavior of complex systems over time. Consequently, it develops mathematical modeling technique for framing, understanding, and resolving the complex issues and problems. What makes using system dynamics different from other approaches to studying complex systems is the use of feedback loops, stocks, and flows. These elements help describe how even seemingly simple systems display baffling nonlinearity. First, SD concepts have been implemented for solving the organizational problems. Subsequently, because of their flexibility in integrating with simulation packages and graphic user interfaces (GUIs), their applications have been widespread. The methodology found application in different problems that varied widely in scope (single organization to national and economies), business process (supply chain management, project management, service delivery, information technology [IT] infrastructure, and strategic planning), and business types (manufacturing, service, research and development, health care, insurance, military, and government).

SD models can portray the structure and behavior of the supply chain systems in greater detail. They can handle a large number of variables, can capture nonlinear relationships, and can model subjective aspects that govern managerial values and their policy decisions. These models take long-term views of the problems and adopt a systems approach to solve these problems. The models are dynamic and depict the evolution of systems in time.

SD models have great explanatory power. This is because SD models are made up of simplifying structures that make them very transparent. The diagramming techniques used in the SD methodology are excellent materials for understanding, conceptualizing, and analyzing models (Wolstenholme, 1992). Causality has been widely acclaimed throughout the ages as the cornerstone of learning. Bagozzi (1980) says that causality is at the heart of human understanding. As SD models are causal models, they add precision to one's theory, make it explicit, allow better representation of complex interrelationships, and formalize scientific inquiry.

SD modeling is a methodology for studying the dynamics of real-world systems. It has its origins in the control-engineering work of Jay Forrester. The central concept of SD is that all objects in a system interact through causal relationships. SD asserts that such relationships form the underlying structure for any system. The creation of a complete dynamic model of a system requires the identification of the causal relationships that form the system's feedback loops. Feedback loops can be either negative, which leads to a goal-seeking behavior, or positive, which can produce positive exponential growth, steady-state, or decline depending upon the gain parameter. From the causal loops, the stock and flow structure is developed. Stocks are accumulations of a physical quantity that characterize the state of the system. Flows are rates that are added to (inflows) or subtracted from (outflows) a stock, and they represent the management policies to control and regulate the state of the system. The stocks and flows mode is then mapped into a mathematical representation (Rabelo et al., 2007).

4.3 Supply Chain Modeling Using System Dynamics

Although the use of SD modeling in the supply chain has been limited, it has gained popularity in recent years. The dynamic nature of supply chain systems and the uncertainties of customer demand, suppliers, logistics routes, and inventory methods have necessitated this change. Forrester (1961) was one of the first to study a theoretical supply chain including retailer, wholesaler, distributor, and manufacturing using SD. The model had shown the inventory fluctuations over time. Sterman (1989) has worked extensively with the well-known Beer Distribution Game and commented on the effects of misperceptions of feedback in decision making.

Over the last decade, more attention has been paid to the use of SD simulation for analyzing different aspects of supply chain management (Ashayeri and Lemmes, 2006). The demonstration of the bullwhip effect is, perhaps, one of the most useful applications of SD in supply chains (Lee at al., 2004; Fiala, 2005).

Utilization of system dynamics in supply chain modeling can be traced back from the work of Forrester. Forrester in 1961 developed industrial dynamics, which he later extended and called SD. In fact, Forrester had already developed a model for the following supply chain, without using the term *supply chain*. His supply chain (which is theoretical, academic) had four links: retailer, wholesaler, distributor, and factory. He examined how these links react to deviations between actual and target inventories. He found that "common sense" strategies may amplify fluctuations in the demand by final customers, up in the supply chain (Kleijnen, 2005). Wikner et al. (1991) compared various methods of improving total dynamic performance of a supply chain using a three-echelon Forrester production distribution system as a supply chain reference model.

A case study was provided by Higuchi and Troutt (2004), who used system dynamics simulations to study the supply chain of short life-cycle products. The model constituted three components of market, retail, and factory and assumed that the diffusion of a new product is based on the logistic curve. System dynamics simulation showed the influence of the supply chain phenomena of boom and bust and the bullwhip effect on the supply chain dynamics. They also analyzed the investment policy by manufacturers under aggressive, neutral, and conservative scenarios using the system dynamics model. Through this, they rejected the existing assertion that the companies control the level of the manufacturing capacity by setting the investment policy. Finally, based on SD model results, it was concluded that it is more important to set the product and supply chain specifications right in the beginning as there might not be enough time to improve them later and reap the benefits under the short life-cycle assumptions.

Spengler and Schroter (2003) used system dynamics to study the supply chain of spare parts of electronic equipment. They modeled an integrated production and recovery system for spare parts supply and evaluated possible strategies for meeting spare-parts demand in the end-of-life service period. They also examined the effect of component recovery in reducing the costs of take-back and product recycling. The model also aimed at examining the behavior of the system in situations when planners underestimate the spare parts demand and decision makers have to meet the customer requirements. Finally the model was utilized for testing alternate policies for managing spare parts in closed-loop supply chains. To sum up, this study aimed at enabling the decision makers in the industries to determine whether and how component recovery could reduce costs in spare parts management in the end-of-life service period. Moreover, the SD model developed was to aid the decision makers in assessing the risks involved with the spare parts management–related policies and in developing policies to improve the dynamic behavior of the system.

Ashayeri et al. (1998) modeled a large European distribution chain using SD. Their paper explains the advantages of using SD in simulating strategic decision problems. From the SD simulations, they concluded that the increase of production capacity, which is a structural change, does not guarantee a stable supply chain. The model results showed that as the lead time increases, the degree of supply chain instability becomes large. Finally the model was used to test three scenarios to select the best business process reengineering strategy.

Minegishi and Thiel (2000) applied SD in modeling a food supply chain. Their study was mainly aimed at showing how SD could be effective in improving the knowledge of the complex logistic behavior of an integrated food industry. The study was done in the poultry production and processing field. The study mainly focused on the problems of coordination between variables controlling various activities in the food supply chain, and on formalizing the cybernetic mechanisms allowing the control of this particular logistic chain. With this aim, an SD model was developed to study the system behavior in order to identify and modify the symptoms of instabilities in these production systems. Various policies relating to the particular food supply chain considered were also analyzed using the developed model.

Ovalle and Marquez (2003) carried out an assessment study on the effectiveness of using e-collaboration tools in the supply chain with SD. The study presented a classification of "managerial spaces" where multiple trading partners share critical information using e-collaboration tools and assess the possible local and global impact on the supply chain performance. This study was facilitated by means of a conceptual supply chain model and was realized using SD simulation. The SD model enabled the assessment of the impact of using e-collaboration tools in a supply chain. The modeling finally led to a conclusion that gradual increments of the information sharing produce positive increases in the local and global performance of the supply chain concerned.

Zhang and Dilts (2004) researched the supply chain network organization structures categorized by the organic and mechanistic management control structures. SD simulation was utilized for studying the structural impacts on cost and fill rate performance in two-echelon and two-supply-chain network organization models under different market coordination conditions. The simulation results proved the effects of demand and network structural factors, and their interactions, on the above measures. From the analysis, the authors came to a conclusion that as the demand becomes dynamic, the cooperative interaction model has a better system performance than the competitive supply chain model.

Sanchan et al. (2004) utilized SD in developing a grain supply chain cost model in the context of India. The study was aimed at developing different models, namely, the cooperative model, the contract farming model, and the collaborative model, in order to minimize the total supply chain cost under optimistic, most likely, and pessimistic scenarios. Nine different scenarios were modeled and evaluated using the SD methodology to generate the future outcome and to devise policies accordingly. The major contribution of the study was in finding the reduced cost ratios in different scenarios. Using this model, various cost elements could be analyzed to get an overall picture of the supply chain cost, thereby identifying the areas for improvement. The model also helped the channel members to understand the system's behavior with respect to various cost elements under different market scenarios.

Reiner (2005) used an integrated approach composed of system dynamics and discrete event simulation to explain the importance of customer orientation for evaluation and improvement of supply chain processes of an electronic manufacturer in the telecom industry. The author showed that it is important to take into account the interdependencies with performance measures in order to integrate the supply chain performance management requirements. The model illustrated the use of performance measures and indicators with the help of a process improvement case study. Through his modeling effort, the author showed how customer satisfaction could impact the financial results.

Narasimha Kamath and Roy (2005) presented a SD-based experimental method to design a supply chain structure of a volatile market of short life-cycle product. They had considered a capacity constrained supply chain in which the product's market acceptance is impeded by limiting product availability. Deciding the strategy for capacity augmentation was considered as one of the major challenges in designing the supply chain. A two-echelon supply chain was considered by the authors. The authors had made use of loop dominance analysis to identify feedback loops that primarily determined the system behavior. They advocated the strengthening of the dominant feedback loop in order to achieve significant improvements in performance.

Georgiadis et al. (2005) adopted SD in modeling and analyzing strategic issues in food supply chains. The modeling was for mapping and analyzing multiechelon food supply chains. The modeling was done mainly to examine the capacity planning policies for a food supply chain with transient flows due to market parameters and constraints. The developed SD model could be utilized to identify realistic policies and optimal parameters for various strategic decision-making problems. The model was successfully implemented for the transportation capacity planning process of a Greek fast-food restaurant supply chain. One of the advantages of the SD model developed was that it could be used to carry out what-if analyses in the context of long-term operation of supply chains using total supply chain profit as the performance measure, and could be of utility to decision makers dealing with various strategic food supply chain management issues.

Kumar and Yamaoka (2005) used SD in studying closed-loop supply chain design in the Japanese car industry. Using the SD model, relationships between reduce, reuse, and disposal in the Japanese car market were examined with a base scenario analysis using the car consumption data and forecast. The model was further utilized in analyzing different market scenarios for the car industry's reverse supply chain. The modeling results showed that the Japanese "end-of-life vehicle program" would trigger the growth of used car exports to emerging countries, and imposing tax on used car export would have some control on such exports and improve economic opportunities for remanufacturers, recyclers, government, manufacturers, and consumers in the country. The model was also used to forecast how various logistics elements would be impacted by government regulations on a long-term basis in the car industry.

Vlachos et al. (2007) used SD in studying product recovery operations in reverse supply chains. The authors developed a SD model for strategic remanufacturing and collection capacity planning of a single product reverse supply chain for product recovery. The developed SD model considered the analysis of the system operations by taking into account the capacity considerations and environmental protection policies on product demand. The model results showed that the decision to develop reverse channel operations must be done using leading capacity expansion strategies. The model also gave insights that can be utilized in developing efficient capacity planning policies in a dynamic manner.

Saeed (2008) studied the use of trend forecasting in determining the ordering policy in supply chains. Trend forecasting is usually used to assess demand, which is a tracked variable in the control context, and which in turn drives supply, a tracking variable. The study involved the construction of a SD model of the classical control mechanisms that involve use of the trend of a tracking variable to improve tracking performance in the context of reducing the bullwhip effect in a supply chain.

Drawing on SD methodology, Georgiadis and Besiou (2008) examined the impact of ecological motivation and technological innovations on the long-term behavior of a single producer, single product closed-loop supply chain with recycling activities. The SD model developed could be used for sensitivity analyses on issues such as the firms' compliance to regulatory measures and green consumerism. The model would aid the decision makers in the management of closed-loop supply chains as well as the researchers in the field of environmental management. The developed model was implemented in a real-world supply chain of electrical equipment in Greece.

Olvera (2009) asserted that supply chain members adopt hybrid business models in order to respond to the changes in the customer demands. He utilized SD to have a quantitative evaluation of the influence of these hybrid business models on the supply chain performance. The model built was used for testing different scenarios to collect total response time and order backlogs. From the modeling it was found that there is a trade-off between total response time and total backlog. Based on the model results, it was finally concluded that the operation of the supply chain is highly affected by the degree of hybrid business models used by the supply chain partners.

Campuzano et al. (2010) integrated SD and fuzzy sets for improving supply chains. A SD model coupled with fuzzy estimations of demand was developed for the simulation of a two-stage, single item, multiperiod supply chain. The fuzzy numbers used in the model were based on the possibility theory to represent demand and orders. Using the SD model, it was shown that the bullwhip effect and the amplification of the inventory variance can be effectively reduced by the fuzzy estimations of the demand.

Fan et al. (2010) studied the bullwhip effect in a military weapons maintenance supply system using SD. SD methodology was used with the aim of simulating the system environment problems such as aging machines, low equipment reliability, weapon life cycle, and inefficient organization, and in developing strategies that can improve the system efficiency and effectiveness in the military supply chain. The research findings indicated that human resource inefficiency was a major problem in the military weapon maintenance supply system, and organizational restructure could be a remedial action to reduce the bullwhip effect and improve performance of the system.

Riddalls and Bennett (2010) analyzed the stability of supply chains using SD. The SD model was developed for the well-known beer game (Sterman, 1989) by considering pure delays rather than exponential lags. Such a consideration led to qualitatively different stability properties. The paper also studied the effect of stock-outs in lower echelons on stability in the supply chain. The authors found that overreaction in amending the routine ordering policy during the adversity periods leads to more instability in the future.

Samuel et al. (2010) analyzed the health service supply chain systems utilizing SD. The SD models considered three service stages. A small increase in the input applications led to amplified variability of backlogs in the subsequent stages. The simulation results showed that reducing capacity adjustment and service delays gave better results in terms of supply chain performance. It was concluded that reducing stages and capacity adjustment were the strategic options for service-oriented supply chains.

Smits (2010) has developed a SD model for supporting the management of intake and treatment processes in the mental health care system. The SD model was used to model the existing system and carry out policy experimentations in order to redesign the intake and treatment processes of the system considered. The author found that shifting of personnel resources between intake and treatment activities did not substantially improve system performance.

Kim and Park (2010) researched linked decision making in a vendor-managed inventory relationship, where the retailer decides the retail price and the vendor determines its capacity commitment to the retailer. The authors had made use of SD simulations based on differential game theory for this purpose. The model considered coordinated decision making of the retailer and the vendor in order to maximize their individual profits. Specific product characteristics such as innovation and imitation effects of demand were also considered in the model. The results showed that that dynamic coordination between the vendor and the retailer on key decision variables is an important consideration.

Springer and Kim (2010) researched the supply chain designs and policies that can minimize supply chain volatility. Using a SD approach, three distinct supply chain volatility metrics were used to evaluate the capability of a static as well as a dynamic policy for a pipeline inventory management situation. A demand shock was introduced to create instability in the system. It was found that the static policy, as prescribed in the SD texts, did not show better results for all three volatility metrics. While the static policy led to a smaller transient bullwhip effect and was less likely to experience endogenous oscillations, the dynamic policy was found to be quicker to converge to a new equilibrium. The dynamic policy also provided fewer stock-outs and backorders.

Barlas and Gunduz (2011) used SD toward investigating the structural sources of the bullwhip effect in supply chain and explored the effectiveness of information sharing to eliminate the undesirable fluctuations. In their model, the authors had considered parameters of certain standard ordering policies, external demand, and lead time. SD simulations ascertained isolated demand forecasting performed at each echelon of the supply chain as a major structural cause of the bullwhip effect. It was also found that demand and forecast sharing strategies can considerably reduce the bullwhip effect.

Hussain and Drake (2011) analyzed the effect of batching on the bullwhip effect in a model of a multiechelon supply chain with information sharing. The authors considered SD simulation as an appropriate methodology to investigate the effect of varying batch size on the bullwhip effect with a stochastic demand process as other methods such as the transform techniques of control theory were found to be extremely challenging. The authors observed that the relationship between batch size and demand amplification was nonlinear in nature. While large batch sizes, combined in integer multiples, produced order rates close to the demand and produced small amplifications in demand, smaller batch sizes led to the greatest value of information sharing and a much greater improvement in the amplification ratio.

Khaji and Shafaei (2011) used a SD approach for studying strategic partnering in supply chain networks. Supply chain coordination and strategic partnering were conceptually integrated in their study. The developed SD model addressed the whole supply chain starting from the suppliers to the final customers including the production and distribution actors. The SD model was generic in nature and was developed to adapt to various network structures. Using this model, different scenarios based on costs and benefits were designed, and the results were analyzed to aid the decision makers in a supply network.

Kumar and Nigmatullin (2011) examined a nonperishable product food supply chain performance under a monopolistic environment using SD. SD was applied with the aim of studying the behavior and relationships in the supply chain and to determine the impact of demand variability and lead time on supply chain performance. The authors claimed that the model can be used to study processes and relationships in the nonperishable food supply chains and can be used as a tool to analyze the design of the supply chain network. The developed SD model was also found useful in studying the relationships between supply chain stages and in analyzing the effect of changing values for the model variables.

Angerhofer and Angelides (2000) and Otto and Kotzab (2003) provide excellent reviews on SD modeling in supply chain management.

4.4 Modeling the Bullwhip Effect Using System Dynamics

4.4.1 The Bullwhip Effect

Uncertainty can alter the effectiveness of a supply chain (Davis, 1993). In a multiechelon supply chain consisting of retailer, wholesaler, and manufacturer, there could be substantial distortions in the downstream information to the members upstream. The distortion can actually be amplified several-fold as it moves upstream, causing the upstream members to make lopsided order and production decisions. When customers order a little more than usual, the retailer orders even more to wholesalers, and the wholesalers order much more to the manufacturers. The manufacturers cannot respond quickly by supplying more as they need to add more capacity and organize additional supplies for producing more. This lack of supply causes the down-stream players to panic and order even more. Eventually when the manufacturers produce all the additional items and send them downstream, there are not as many takers. The orders get dried up and the inventories pile—the supply chain feels the effect of lack of orders, and the manufacturers decide to cut back in capacity and supplies causing the production to reduce. The cycle continues.

This phenomenon is called the *bullwhip effect* in supply chains (also known as the *whiplash effect*). The name is derived from the fact that a small varia-tion in customer demand can actually cause the upstream players to experi-ence high variances in demand and supply leading to capacity augmentation and downsizing cycles. The bullwhip effect can lead the upstream players to have excessive investment, poor customer service, lost revenues, misguided capacity plans, ineffective transportation, and missed production schedules.

Utilization of SD in modeling the bullwhip effect could be traced back to the work of Forrester (1961). Forrester had developed a theoretical supply chain model (without calling it a "supply chain") including retailer, wholesaler, distributor, and factory to examine how the individual players react to devia-tions between actual and target inventories. He found that "common sense" strategies may amplify fluctuations in the demand by final customers, up in the supply chain (Kleijnen, 2005). Thereafter, one of the best illustrations of the bullwhip effect was found in the well-known "beer distribution game" (popularized by Sterman, 1989) where the players made ordering decisions within a simple supply chain subject to information and shipment lags. In this game, teams of participants managed different levels of the distribution chain: retailer, wholesaler, distributor, and factory. The players at each level received shipments of beer from their suppliers each week, filled as many of their customers' orders as possible from the inventory, and placed new orders for beer with their supplier. The game illustrated the bullwhip effect, and the participants became aware of the fact that the bullwhip effect was arising out of their own decisions. Sterman (1989) also interpreted that the bullwhip effect was the result of incorrect decision making of the players as they were subjected to misperceptions of feedback.

Lee et al. (1997) had identified the bullwhip effect as an outcome of the strategic interactions among rational supply chain members. They had identified four causes of the bullwhip effect: rationing or shortage gaming, demand signal processing, order batching, and price fluctuations. It is possi-ble to counteract the bullwhip effect with a thorough understanding of these causes (Cachon, 1999; Cachon and Lariviere, 1999; Chen et al., 2000a).

Samuel (2005) had studied the dynamics of a multiechelon supply chain, the causes of the bullwhip effect, and the impact of each cause on the bullwhip effect. The research work also focused on the measures that could be taken up to reduce the bullwhip effect. The study made relative comparison of each of the causes of the bullwhip effect in numerical terms and measured the effect of the sharing of demand information on the supply chain performance.

Various authors had listed counteractive measures to alleviate the ills of the bullwhip effect (Gallego and Ozer, 2001; Gavirneni et al., 1999; Lee and Whang, 1998). The measures include the following: time compression, adjustments in ordering policies, improved communications and information sharing, demand visibility, channel integration, and improvements in the operation efficiency.

4.4.2 Modeling the Bullwhip Effect

In this section, an attempt is made to model a decentralized three-echelon supply chain consisting of three sectors: the retailer, the wholesaler, and the manufacturer. Each of the sectors maintains an inventory and an order backlog. A rational ordering policy based on anchoring and adjustment is assumed for each of the supply chain members. Based on the demand from the customer side, the retailer places orders to the wholesaler and the wholesaler in turn places orders to the manufacturer. The replenishment of the orders takes place after a delay period. The supply chain under consideration assumes flow of material from manufacturer to wholesaler, wholesaler to retailer, and retailer to end customer. Figure 4.1 shows the flow diagram of a typical sector (retailer).

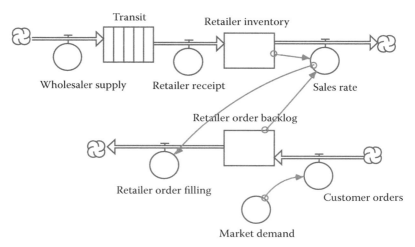

FIGURE 4.1
A typical sector in the base model.

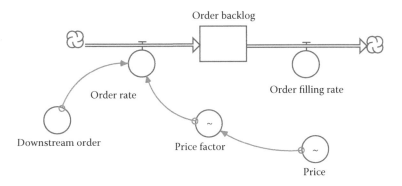

FIGURE 4.2
Modeling of price fluctuations.

4.4.3 Impact of Price Fluctuations on Supply Chains

Price fluctuation is one of the major causes of the bullwhip effect. Price fluctuations that may occur due to discounts, cost flare-ups, or revised tax situations, are enough to send waves of demand shock through the supply chain resulting in high inventory levels or stock-outs. Such fluctuations in orders received by an intermediary create much larger fluctuations in order quantities upstream in a decentralized multiechelon supply chain. Figure 4.2 shows the modeling of price fluctuations for one of the intermediate sectors.

In this model, price is shown as a graph function of time. Price factor, in turn, is a graph function of price. As price increases, price factor decreases, signifying a decrease in the motivation to purchase. Because the order placed is a function of price factor and normal orders, a decreasing price factor leads to less than normal incoming orders.

4.4.4 Impact of Shortage Gaming

Shortage gaming is another major cause of the bullwhip effect. Intermediaries in a decentralized multiechelon supply chain tend to order more than they really require to an upstream member owing to a fear of short supply or rationing. Such a practice creates imbalances in the supply chain.

One way of modeling this situation is by creating a shortage gaming factor and obtaining the order rate as a multiplication of the downstream order with this factor. The shortage gaming factor may be defined as a ratio of the downstream order divided by a forecast of the likely orders to be filled. Figure 4.3 shows the modeling of shortage gaming.

Thus,

Shortage Gaming Factor = (Downstream Order/Order Filling Forecast)

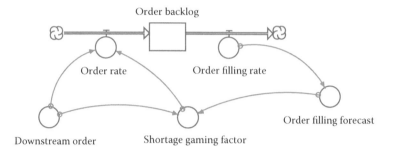

FIGURE 4.3
Modeling of shortage gaming.

This factor can have a maximum value of 1.
  Now,

$$\text{Order Rate} = \text{Downstream Order} \times \text{Shortage Gaming Factor}$$

4.4.5 Demand Signal Processing

Demand signal processing involves updating demand forecasts from downstream level to upstream level in a supply chain based on the increase or decrease of the demand signal. Chen et al. (2000a) studied the effect of demand forecasting and order lead times on the bullwhip effect. They had shown that the bullwhip effect can be reduced but not completely eliminated by centralizing demand information. Chen et al. (2000b) demonstrated that the use of an exponential smoothing forecast by the retailer can cause a bullwhip effect and contrasted these results with the increase in variability due to the use of a moving average forecast. A managerial insight from their research was that the bullwhip effect exists, in part, due to the retailer's need to estimate the mean and variance of demand. The increase in variability is an increasing function of the lead time. The more complicated the demand models and the forecasting techniques are, the greater is the increase. Centralized demand information can reduce the bullwhip effect but will not eliminate it.

  Demand signal processing can be modeled for a decentralized supply chain intermediary by making use of downstream order forecast considering the past values of downstream order along with the usual inventory and order backlog corrections. Thus,

$$\text{Ordering Decision} = \text{Downstream Order Forecast} +$$
$$\text{Inventory Corrections} + \text{Order Backlog Corrections}$$

Figure 4.4 shows the modeling of demand forecast updating.

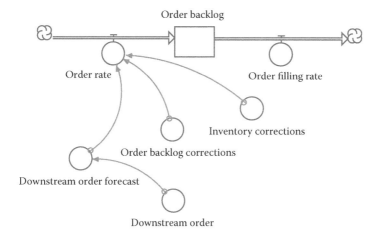

FIGURE 4.4
Modeling of demand forecast updating.

4.4.6 Order Batching

Companies resort to batching of orders to reduce the costs of ordering. Such batching, although beneficial in the short term, may actually lead to bullwhip effects. This is because the upstream suppliers, at times, misestimate the market and get ready for orders that are not going to realize.

The batching of orders in the multiechelon supply chain is modeled with the help of a release counter. The downstream orders keep accumulating until the release counter reaches a specified value. At that time, the order batching stops, and the accumulated downstream orders are released as order rate to the intermediary sector. The release counter is simultaneously reset to a zero value, and the orders start accumulating again.

As an example, if order batching is introduced with the orders allowed to accumulate for a period of 3 months, then the release counter will start from a value of 0, rise to a value of 3 when it is reset to a zero value, and simultaneously all the accumulated orders will be released. Figure 4.5 shows the flow diagram of the relevant portion of the model.

4.5 A Case of Retailer's Inventory

A retailer stocks a consumer good in inventory and ships the good to customers based on their constant demand of 20 units. The retailer follows a policy of immediately ordering exactly the same quantity to its supplier that is

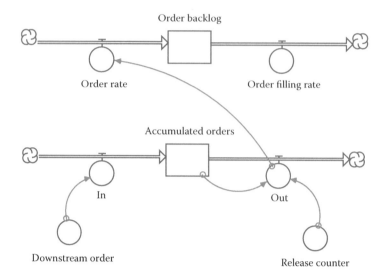

FIGURE 4.5
Modeling of order batching.

shipped to the customers. The supply is prompt and with a constant delivery lead time of exactly 8 days for all the replenishment orders from the retailer.

The initial inventory held is 200 units and the quantity of order backlog with the supplier is 100 units. The supply chain remains in a steady state for the first 5 days (i.e., the inventory and order backlog maintain their steady-state values of 200 and 100, respectively). Figure 4.6 shows this situation.

Now assume that after the first 5 days, there is a sudden step rise in customer demand to 25 units per day and then it remains constant. Figure 4.7 shows the step rise in customer demand.

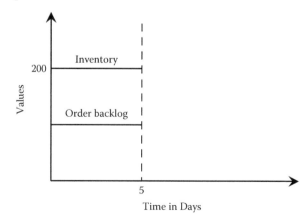

FIGURE 4.6
Initial steady-state values of inventory and order backlog.

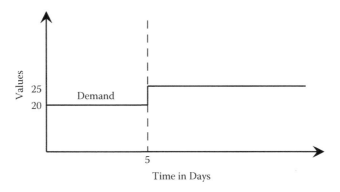

FIGURE 4.7
Step rise in customer demand.

What would be the behavior of the retailer's inventory and order backlog after the step rise in demand? The answer to the question could be found by SD modeling using the Stella software. Figure 4.8 shows the flow diagram of the retailer's inventory situation. The SD equations are given thereafter. Finally, Figure 4.9 shows the behavior of inventory and order backlog after the step rise in demand.

The SD equations could be written as follows:

```
Order_Backlog(t) = Order_Backlog(t - dt) + (Order_Rate -
    Supply_Rate) * dt
INIT Order_Backlog = 100
INFLOWS:
Order_Rate = Shipping_Rate
OUTFLOWS:
Supply_Rate = DELAY(Order_Rate, Supply_Delay)
Retailer_Inventory(t) = Retailer_Inventory(t - dt) + (Supply_
    Rate - Shipping_Rate) * dt
INIT Retailer_Inventory = 200
INFLOWS:
Supply_Rate = DELAY(Order_Rate, Supply_Delay)
OUTFLOWS:
Shipping_Rate = 20+STEP(5,5)
Supply_Delay = 8
```

Figure 4.9 shows that retailer's inventory decreases steadily from its initial value of 200 for the next 8 weeks (the period of constant delivery lead time) and reaches a new steady-state value of 160. On the other hand, order backlog increases from its initial value of 100 to a new steady-state value of 140 over a period of 8 weeks. It is interesting to note that although the retailer makes a replenishment decision of ordering exactly the same quantity to its supplier that is shipped to a customer, a step rise in the customer demand leads to

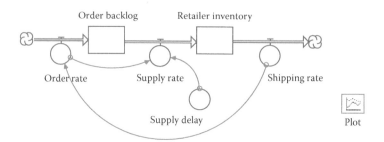

FIGURE 4.8
Flow diagram of the retailer's inventory situation.

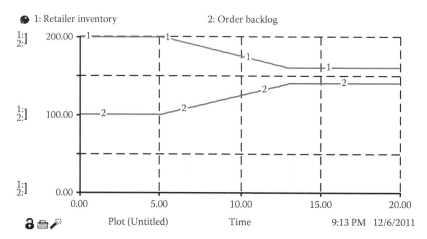

FIGURE 4.9
Inventory and order backlog behavior.

a lower steady-state value of the retailer's inventory. The retailer was earlier maintaining a steady-state inventory of 10 days of customer demand (inventory 200 units with a customer demand of 20 units). In the new steady state, the retailer is maintaining an inventory that is slightly more than 6 days of customer demand (inventory 160 units with a customer demand of 25 units).

The case study shows that the retailer needs a better policy of ordering replenishments in order to maintain sufficient inventory. The policy of placing a replenishment order of exactly the same quantity that is shipped to the customers is not working well for the retailer.

4.6 Conclusions

SD modeling is increasingly being used for modeling supply chains. Such modeling exercises can benefit from the explanatory power of system dynamics. On the one hand these models can depict the cause-and-effect connections to the researchers, while on the other hand, the simulation results can be of help in making useful conclusions. The models are also useful for carrying out policy experimentations and in recommending policies and what-if analyses to the decision makers.

After giving a broad description of SD and its relevance to supply chain modeling, this chapter presented a brief review of the relevant literature. A section is devoted to developing models of the bullwhip effect, and four major causes of the bullwhip effect are modeled with the help of SD in a decentralized supply chain setting. Finally, a case of retailer's inventory is discussed in detail.

Appendix: A Brief Note on System Dynamics

System Dynamics Concepts

SD was developed in the late 1950s by Jay W. Forrester at the Sloan School of Management, Massachusetts Institute of Technology (MIT). SD encourages holistic thinking based on the application of control engineering theory to management, and based on closed-loop feedback thinking. According to Wolstenholme (1990), SD is a rigorous method for the qualitative description, exploration, and analysis of complex systems in terms of their processes, information, organizational boundaries, and strategies, which facilitates quantitative simulation modeling and analysis for the design of system structure and behavior.

According to Coyle (1977), causality is characterized in the following ways:

- Conservation considerations
- Direct observation
- Instruction to that effect
- Accepted theory
- Hypothesis/assumptions or beliefs
- Statistical evidence

Out of these, the conservation considerations are made when a causal variable accumulates in time to produce the affected variable. Such considerations always occur in a physical flow of people, material, cash, equipment, or orders. For example, birth accumulates to population, production accumulates to inventory, and order accumulates to order backlog. Death also accumulates to population, but in a negative way, so is depreciation to capacity, sales to inventory, and so on. Thus, we have

- Birth → + Population
- Production → + Inventory
- Order → + Order Backlog
- Death → - Population
- Depreciation → - Capacity
- Sales → - Inventory

The direction in each case is determined by assuming a steady-state condition and observing the direction of change in the affected variable for a change in the value of the causal variable. Other causal characterizations usually take place in information flow. Say, for example, with more quality, more orders are received. This may be shown as

- Quality → + Order

Similarly, the more the productivity, the more is the production; the more the capacity, the more is the maintenance cost. On the other hand, the more the price, the less is the demand; the more the fatigue, the less is the output; and so on:

- Productivity → + Production
- Capacity → + Maintenance Cost
- Price → - Demand
- Fatigue → - Output

In all the examples above, causal links exist between two variables. Such causal links in series give rise to causal chains. Here are some examples:

1. Quality → + Order → + Order Backlog → + Delivery Delay
2. Productivity → + Production → + Inventory → + Carrying Costs

By seeing all the examples above, one should not get an idea that each affected variable is influenced by only one causal variable. In fact, more often than not, multivariate causality is present. For example, see Figure 4.10.

There are two axioms in SD. These can be stated as follows:

- Feedback loops constitute the structure of a system.
- System behavior is a function of the system structure and the policies.

A feedback loop results because of a closed sequence formed by a causal chain. As the first axiom states, it is the basic unit that forms the structure

FIGURE 4.10
Multivariate causality.

of a system. The second axiom states that system behavior results from its structure. Thus, feedback loops present in a system determine its behavior.

Policies are decisions that can alter the behavior of a system. Such decisions may actually alter the feedback loops present in a system or make important changes in them. The main purpose of SD modeling is to make observations about the behavior of a system based on existing as well as new policies.

There are two types of feedback loops: positive and negative. A positive loop can show positive exponential growth, decline, or steady behavior. A negative loop, on the other hand, shows goal-seeking behavior, guided by the negative exponential distribution. A higher-order negative feedback loop system also shows oscillations. There are some excellent texts on SD by Forrester (1961), Goodman (1974), Mohapatra et al. (1994), and Sterman (2000).

Two diagrams are used in SD to depict a SD model: the causal loop diagram and the flow diagram. While the causal loop diagrams depict cause-and-effect sequences, the flow diagram shows physical and information flows.

The rest of this appendix gives three simple examples of SD applications. The first two examples are on asset value predictions: one on a positive feedback loop situation and the other on a negative feedback loop situation. The third example shows SD modeling of a simple production-inventory system.

EXAMPLE 1

Asset Value Prediction—Positive Feedback Loop Situation

A person buys some assets at year zero. The asset value appreciates in a manner that 10% of the available assets are added every year. The person sells a fixed number of assets every year. What would be the asset value in the next 12 years?

In order to model this situation, a flow and a causal loop diagram are drawn first. (See Figure 4.11.)

Because the model has only one positive feedback loop, it will have three possible behavior modes: positive exponential growth, steady state, and positive exponential decline. These behavior modes are shown in Figure 4.12.

The exact behavior of a system will depend on the relative values of *In Factor* and the fixed outflow *Out*.

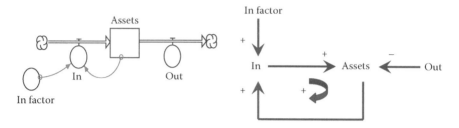

FIGURE 4.11
Positive feedback loop situation: flow diagram and causal loop diagram.

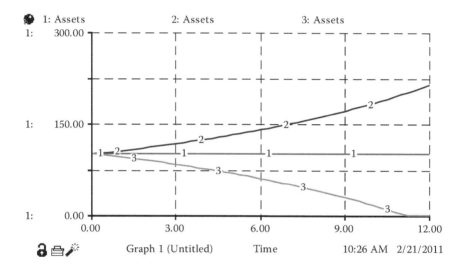

FIGURE 4.12
The three behavior modes in a positive feedback loop (1, steady state; 2, positive exponential growth; and 3, positive exponential decline).

EXAMPLE 2

Asset Value Prediction—Negative Feedback Loop Situation

A person buys some assets at year zero. The asset value appreciates at a constant value every year. The person sells 10% of the available assets every year. What would be the asset value in the next 12 years?

In order to model this situation, a flow and a causal loop diagram are drawn first. (See Figure 4.13.)

Because the model has only one negative feedback loop of first order, it will have three possible behavior modes: negative exponential growth, steady state, and negative exponential decay. These behavior modes are shown in Figure 4.14.

The exact behavior of a system will depend on the relative values of *Out Factor* and the fixed inflow *In*.

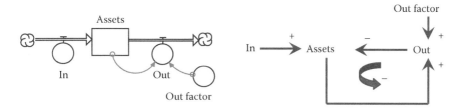

FIGURE 4.13
Negative feedback loop situation: flow diagram and causal loop diagram.

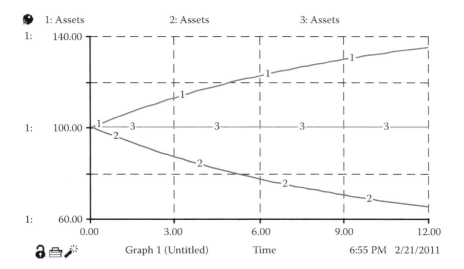

FIGURE 4.14
The three behavior modes in a negative feedback loop (1, negative exponential growth; 2, negative exponential decay; and 3, steady state).

EXAMPLE 3

A Simple Production Inventory System

Let us consider a simple production inventory system. The procurement of materials accumulates to an inventory. The inventory is depleted by the consumption of material. The situation can be depicted by an initial flow diagram as shown in Figure 4.15.

In this flow diagram, *Inventory* is a level variable and *Procurement Rate* and *Consumption Rate* are rate variables. Level variables are those that can be measured at an instant of time and are essentially accumulations of a physical quantity such as materials. The rate variables, on the other hand, can only be measured in a period of time and depict the flow of the physical quantity.

Now suppose the *Consumption Rate* is exogenous and the *Procurement Rate* is controllable. It is planned to build up a *Desired Inventory*, and to

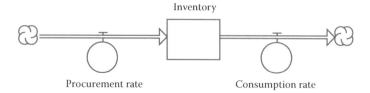

FIGURE 4.15
Initial flow diagram of the simple production-inventory system.

obtain this, the *Procurement Rate* is modeled as an addition of *Consumption Rate* as an anchor and *Discrepancy* from the desired inventory divided by the *Adjustment Time* as an adjustment. It is an assumption here that any material ordered is obtained immediately without any loss of time. The complete flow diagram is presented in Figure 4.16.

In this diagram, *Inventory* is a level variable, *Procurement Rate* and *Consumption Rate* are rate variables, *Discrepancy* is an auxiliary variable, and *Adjustment Time* and *Desired Inventory* are constants.

The causal-loop diagram is shown in Figure 4.17.

The causal loop diagram depicts the cause-and-effect relationships in the model. As can be observed, the direction of arrows in the flow diagram and causal loop diagram are not all the same. In fact, for *Consumption Rate* to *Inventory*, causal direction (as in cause–effect diagram) is exactly opposite the flow direction (as in flow diagram). In all other cases, the directions are the same. Why is this so? Actually, *Consumption Rate* is an outflow that depletes *Inventory*. *Consumption Rate* is a cause, and *Inventory* is an effect just like *Procurement Rate* is a cause and *Inventory* is an effect. Hence, in the causal loop diagram, the direction of the arrow is from *Procurement Rate* to *Inventory* and from *Consumption Rate* to *Inventory*.

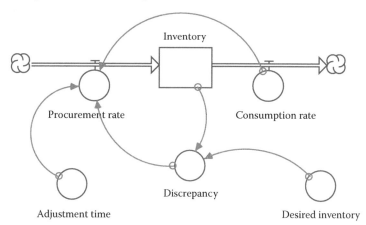

FIGURE 4.16
Complete flow diagram of the simple production-inventory system.

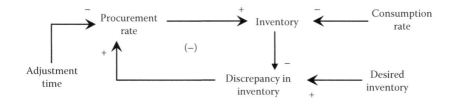

FIGURE 4.17
Causal loop diagram of the simple production-inventory system.

The causal-loop diagram also shows a feedback loop. The feedback loop is from *Procurement Rate* to *Inventory* to *Discrepancy* to *Procurement Rate*. The feedback loop is given a negative sign because the number of negative signs in this feedback loop is odd (one in this case). A negative feedback loop is goal-seeking. In this case, it seeks to make *Inventory* equal to the *Desired Inventory*. The SD equations for the model are given below:

> *Constants*
>> Desired inventory: $DINV = 400$
>> Adjustment time: $AT = 10$

> *Initial values*
>> Inventory: $INV(0) = 200$
>> Initial consumption rate: $CR(0) = 5$

> *Auxiliary variable*
>> Discrepancy in inventory: $DISC(t) = DINV - INV(t)$

> *Rate variables*
>> Procurement rate: $PR(t) = DISC(t)/AT$
>> Consumption rate: $CR(t) = CR$

> *Level variable*
> Inventory: $INV(t + DT) = INV(t) + DT * (PR(t) - CR(t))$

With the initial values and constants as given above, the simulation run, as depicted in Figure 4.18, shows the goal-seeking behavior of the level variable *Inventory*.

The desired inventory is having a value of 400. The inventory plot with time seeks this value of desired inventory and eventually almost reaches it after about 40 time units. Thereafter, inventory remains steady at that value.

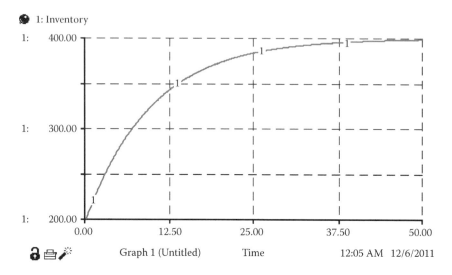

FIGURE 4.18
Simulation run—goal-seeking behavior of *Inventory*.

References

Angerhofer, B.J., and Angelides, M.C. 2000. System dynamics modeling in supply chain management: Research review. In J.A. Joines, R.R. Barton, K. Kang, and P.A. Fishwick (Eds.), *Proceedings of the 2000 Winter Simulation Conference*. 342–352.

Ashayeri, J., Keij, R., and Broker, A. 1998. Global business process re-engineering: A system dynamics based approach. *International Journal of Operations and Production Management*, 18, 817–831.

Ashayeri, J., and Lemmes, L. 2006. Economic value added of supply chain demand planning: A system dynamics simulation. *Robotics and Computer-Integrated Manufacturing*, 22, 550–556.

Bagozzi, R.P. 1980. *Causal Models in Marketing*. New York: Wiley.

Barlas, Y., and Gunduz, B. 2011. Demand forecasting and sharing strategies to reduce fluctuations and the bullwhip effect in supply chains. *Journal of the Operational Research Society*, 62, 458–473.

Beamon, B.M. 1998. Supply chain design and analysis: Models and methods. *International Journal of Production Economics*, 55, 281–294.

Cachon, G.P. 1999. Managing supply chain demand variability with scheduled ordering policies. *Management Science*, 45(6), 843–856.

Cachon, G.P., and Lariviere, M.A. 1999. Capacity choice and allocation: Strategic behavior and supply chain performance. *Management Science*, 45(8), 1091–1108.

Campuzano, F., Mula, J., and Peidro, D. 2010. Fuzzy estimations and system dynamics for improving supply chains. *Fuzzy Sets and Systems*, 161, 1530–1542.

Campuzano, F., and Mula, J. 2011. *Supply Chain Simulation: A System Dynamics Approach for Improving Performance*. London: Springer-Verlag.

Chen, F., Drezner, Z., Ryan, J.K., and Simchi-Levi, D. 2000a. Quantifying the bullwhip effect in a simple supply chain: The impact of forecasting, lead times, and information. *Management Science*, 46(3), 436–443.

Chen, F., Ryan, J.K., and Simchi-Levi, D. 2000b. The impact of exponential smoothing forecasts on the bullwhip effect. *Naval Research Logistics*, 47(3), 269–286.

Coyle, R.G. 1977. *Management System Dynamics*. Wiley, Chichester, UK.

Davis, T. 1993, Summer. Effective supply chain management. *Sloan Management Review*, 35–46.

Fan, C.Y., Fan, P.S., and Chang, P.C. 2010. A system dynamics modeling approach for a military weapon maintenance supply system. *International Journal of Production Economics*, 128, 457–469.

Fiala, P. 2005. Information sharing in supply chains. *Omega*, 33, 419–423.

Forrester, J.W. 1961. *Industrial Dynamics*. Cambridge, MA: The MIT Press.

Gallego, G., and Ozer, O. 2001. Replenishment decisions with advance demand information. *Management Science*, 47(10), 1344–1360.

Gavirneni, S., Kapuscinski, R., and Tayur, S. 1999. Value of information in capacitated supply chains. *Management Science*, 45(1), 16–24.

Georgiadis, P., and Besiou, M. 2008. Sustainability in electrical and electronic equipment closed-loop supply chains: A system dynamics approach. *Journal of Cleaner Production*, 16, 1665–1678.

Georgiadis, P., Vlachos, D., and Iakovou, E. 2005. A system dynamics modeling framework for the strategic supply chain management of food chains. *Journal of Food Engineering*, 70, 351–364.

Goodman, M.R. 1974. *Study Notes in System Dynamics*. Cambridge, MA: The MIT Press.

Higuchi, T., and Troutt, M.D. 2004. Dynamic simulation of the supply chain for a short life cycle product—lessons from the Tamagotchi case. *Computers and Operations Research*, 31, 1097–1114.

Hogarth, R.M. 1980. *Judgment and Choice: The Psychology of Decision*. Wiley, Chichester, UK.

Hussain, M., and Drake, P.R. 2011. Analysis of the bullwhip effect with order batching in multiechelon supply chains. *International Journal of Physical Distribution and Logistics Management*, 41(8), 797–814.

Khaji, M.R., and Shafaei, R. 2011. A system dynamics approach for strategic partnering in supply networks. *International Journal of Computer Integrated Manufacturing*, 24(2), 106–125.

Kim, B., and Park, C. 2010. Coordinating decisions by supply chain partners in a vendor-managed inventory relationship. *Journal of Manufacturing Systems*, 29, 71–80.

Kleijnen, J.P.C. 2005. Supply chain simulation tools and techniques: A survey. *International Journal of Simulation and Process Modeling*, 1(1/2), 82–89.

Kumar, S., and Nigmatullin, A. 2011. A system dynamics analysis of food supply chains—Case study with non-perishable products. *Simulation Modeling Practice and Theory*, 19, 2151–2168.

Kumar, S., and Yamaoka, T. 2005. System dynamics study of the Japanese automotive industry closed loop supply chain. *Journal of Manufacturing Technology Management*, 18(2), 115–138.

Lee, H.L. 2002. Aligning supply chain strategies with product uncertainties. *California Management Review*, 44(3), 105–119.

Lee, H.L., Padmanabhan, V., and Whang, S. 1997. Information distortion in a supply chain: The bullwhip effect. *Management Science*, 43(4), 546–558.

Lee, H.L., Padmanabhan, V., and Whang, S. 2004. Information distortion in a supply chain: The bullwhip effect. *Management Science*, 50(12), 1875–1886.

Lee, H., and Whang, S. 1998. Information sharing in a supply chain, Research paper series, Research paper no.1549, GSB, Stanford University, 1–19.

Minegishi, S., and Thiel, D. 2000. System dynamics modeling and simulation of a particular food supply chain. *Simulation Practice and Theory*, 8, 321–339.

Mohapatra, P.K.J., Mandal, P., and Bora, M.C. 1994. *Introduction to System Dynamics Modeling*, first ed. India: Universities Press.

Narasimha Kamath, B., and Roy, R. 2005. Supply chain structure design for a short life cycle product: A loop dominance based analysis. In *Proceedings of the 38th Annual Hawaii International Conference on System Sciences* (HICSS'05). 3, 78a, Track 3. Big Island, Hawaii, January 3–6. ISBN: 0-7695-2268-8.

Olvera, C.M. 2009. Benefits of using hybrid business models within a supply chain. *International Journal of Production Economics*, 120, 501–511.

Otto, A., and Kotzab, H. 2003. Does supply chain management really pay? Six perspectives to measure the performance of managing a supply chain. *European Journal of Operational Research*, 144, 306–320.

Ovalle, O.R., and Marquez, A.C. 2003. The effectiveness of using e-collaboration tools in the supply chain: An assessment study with system dynamics. *Journal of Purchasing and Supply Management*, 9, 151–163.

Rabelo, L., Helal, M., Lertpattarapong, C., Moraga, R., and Sarmiento, A. 2007. Using system dynamics, neural nets, and eigenvalues to analyse supply chain behaviour: A case study. *International Journal of Production Research*, 46(1), 51–71.

Reiner, G. 2005. Customer-oriented improvement and evaluation of supply chain processes supported by simulation models. *International Journal of Production Economics*, 96, 381–395.

Riddalls, C.E., and Bennett, S. 2010. The stability of supply chains. *International Journal of Production Research*, 40(2), 459–475.

Sarimveis, H., Patrinos, P., Tarantilis, C.D., and Kiranoudis, C.T. 2008. Dynamic modeling and control of supply chain systems: A review. *Computers and Operations Research*, 35, 3530–3561.

Saeed, K. 2008. Trend forecasting for stability in supply chains. *Journal of Business Research*, 61, 1113–1124.

Samuel, C., Gonapa, K., Chaudhary, P.K., and Mishra, A. 2010. Supply chain dynamics in health care services. *International Journal of Health Care Quality Assurance*, 23(7), 631–642.

Samuel, C. 2005. A system dynamics study of the bullwhip effect in supply chains. Unpublished PhD thesis, IIT Kharagpur, India.

Sanchan, A., Sahay, B.S., and Sharma, D. 2004. Developing Indian grain supply chain cost model: A system dynamics approach. *International Journal of Productivity and Performance Management*, 54(3), 187–205.

Smits, M. 2010. Impact of policy and process design on the performance of intake and treatment processes in mental health care: A system dynamics case study. *Journal of the Operational Research Society*, 61, 1437–1445.

Spengler, T., and Schroter, M. 2003. Strategic management of spare parts in closed-loop supply chains: A system dynamics approach. *Interfaces*, 33(6), 7–17.

Springer, M., and Kim, I. 2010. Managing the order pipeline to reduce supply chain volatility. *European Journal of Operational Research*, 203, 380–392.

Sterman, J.D. 1989. Misperceptions of feedback in dynamic decision making. Organizational Behavior and Human Decision Processes, 43, 301–335.

Sterman, J.D. 2000. *Business Dynamics: Systems Thinking and Modeling for a Complex World*. New York: McGraw-Hill.

Vlachos, D., Georgiadis, P., and Iakovou, E. 2007, A system dynamics model for dynamic capacity planning of remanufacturing in closed-loop supply chains. *Computers and Operations Research*, 34, 367–394.

Wikner, J., Towill, D.R., and Naim, M. 1991. Smoothing supply chain dynamics. *International Journal of Production Economics*, 22(3), 231–248.

Wolstenholme, E.F. 1990. *System Enquiry: A System Dynamics Approach*. Chichester, UK: Wiley.

Wolstenholme, E.F. 1992. The definition and application of a stepwise approach to model conceptualisation and analysis. *European Journal of Operational Research*, 59(1), 123–136.

Wolstenholme, E.F., and Coyle, R.G. 1983. The development of system dynamics as a methodology for system description and qualitative analysis. *Journal of Operational Research Society*, 34, 569–581.

Zhang, Y., and Dilts, D. 2004. System dynamics of supply chain network organization structure. *Information Systems and e-Business Management*, 2, 187–206.

5

Technologies for Supply Chain
Integration and Interoperability

5.1 Introduction

Most supply chain management strategies call for robust coordination and collaboration of supply chain partners. Such coordination is possible through effective information sharing through information and communication technologies (ICTs) in general and the Internet in particular. ICTs have enabled companies to disseminate information and coordinate with upstream as well as downstream partners in real time on various issues such as product design and development, procurement, production, inventory, distribution, after-sales service support, and marketing. The companies are now able to explore new supplier bases and directly communicate with the customers in faster and secure ways through the Internet. The other benefits of information sharing include better asset utilization, faster time to market, reduction in total order fulfillment times, enhanced customer service and responsiveness, penetration of new markets, higher return on assets, use of market intelligence for better demand management, and ultimately, higher shareholder value. Therefore, a thorough understanding of the ICTs can help companies to derive maximum benefits out of these technologies.

Supply chain integration is defined as the extent to which the firm can strategically collaborate with its supply chain partners and collaboratively manage the intra- and interorganization processes to achieve the effective and efficient flows of products and services, information, money, and decisions with the objective of providing the maximum value to the customer at low cost and high speed (Yi-nan, Zhao-fang, 2009). According to Lee (2000), supply chain integration has three aspects: (1) information integration, (2) coordination and resource sharing, and (3) organizational relationship linkages. While the first aspect deals with technical issues involved in integrating heterogeneous information systems, the last two issues are strategic in nature. Information integration is not effective if these strategic issues are not adequately addressed.

The major concerns for integration are efficient business process integration, increased flexibility throughout the company to accept the process integration, and *interoperability* of ICT solutions, systems, and people to face the resulting variability in the business environment (Vernadat, 2002). Interoperability is the ability of two or more systems or components to exchange and to use the information that has been exchanged (IEEE, 1990). Two integrated systems are inevitably interoperable, but two interoperable systems are not necessarily integrated (Chen et al., 2008). Interoperable systems require both technical- and semantics-level interoperability. Technical interoperability deals with hardware and software compatibility of the interconnected systems, and semantic interoperability ensures that both systems have the same understanding of different concepts. Supply chain interoperability for business-to-business (B2B) integration is a particularly challenging task because of factors such as diverse information formats, large and dynamic information space, lack of standards for semantic integration of data, and issues related to fast, secure, and reliable data transmission.

This chapter is organized as follows: we start with a motivating example to understand the complicacies involved in information system integration. Next, we discuss the primary and supporting technologies for supply chain integration. The primary technologies we discuss are electronic data interchange (EDI) and Web services. Many companies started using EDI in the late 1960s. During this period, adoption of EDI used to be a costly affair with the use of enormous computer systems and proprietary networks. Recently, with the Internet and data exchange standards such as XML, EDI implementation has become cost effective. Web service is the latest technology for enterprise integration. EDI integrates business processes at the data layers, and Web services allow integration of workflows at the process level. The supporting technologies that we discuss include security and payment systems, and automatic product data capture technologies like radio frequency identification (RFID). In each case we have four sections: overview of the underlying technology, existing standards, management issues, and research directions. Finally, we discuss the technology impact on two business functions that bind an organization with its immediate upstream and downstream supply chain partners: procurement and customer relationship management.

5.2 Dimensions of Supply Chain Integration

According to Lee and Whang (2001), supply chain integration has four dimensions: information integration, planning synchronization, workflow coordination, and new business models and monitoring. Zeng and Pathak (2003) on the other hand suggest the dimensions to be customer/market integration, information integration, logistics and distribution integration, and supplier integration and purchasing integration. Kannan and Tan (2010) support these ideas and suggest four dimensions of supply chain integration: customer focus, supplier focus, supply chain focus, and information flow focus. Comparing the work of Lee and Whang (2001), one can find that the four dimensions proposed by them are individually applicable to customer, supplier, and service providers. We combine both the ideas and perceive that integration is possible in each stage of the supply chain: with upstream members (customer focus), with downstream members (supplier focus), and while managing internal processes with outsourced service and manufacturing operations (value chain focus). Contract manufacturer and 3PL are examples of such outsourcing service providers. We call these three aspects *supply chain dimensions*. Each supply chain dimension is associated with four *information exchange* dimensions suggested by Lee and Whang (2001). Table 5.1 shows these dimensions with some example of business processes in each dimension.

TABLE 5.1

Dimensions of Supply Chain Integration

Supply Chain Information Exchange	Example Business Processes		
	Upstream (Supplier Focus)	Internal/Service Providers (Value Chain Focus)	Downstream (Customer Focus)
Information integration	Point of service (POS) information sharing	Design data sharing and plan sharing	Customer usage data sharing
Planning synchronization	Coordinated replenishment	Synchronized new product introduction and rollover plan, demand–supply management and collaborative planning	Service supply chain planning and coordination
Workflow coordination	e-Procurement, reverse auction, auto replenishment, auto payment	Workflow automation with logistics providers and contract manufacturers, product data management, and collaborative design	Auto replenishment, dynamic pricing
New business models	Business-to-business market exchanges	Spend analysis and management, knowledge management	Market intelligence, mass customization, and new service offering
Monitoring and measurement	Contract agreement compliance monitoring, supplier evaluation, and scorecarding	Logistics tracking, order monitoring, project monitoring	Performance measurement and monitoring

Information integration refers to the sharing of information among members along the supply chain. For example, a firm may share demand forecast and inventory information with the upstream members, promotion plan with downstream partners, and product design and planning data with the contract manufacturers. *Planning synchronization* refers to the joint design and execution of plans for forecasting and replenishment. *Workflow coordination* refers to highly streamlined workflow activities between supply chain partners. For example, procurement activities from a manufacturer to a supplier can be tightly coupled through an e-procurement system so that efficiencies in terms of accuracy, time, and cost can be achieved. *New business models* refer to new ways of doing business in a supply chain. One example is B2B markets that help discover new suppliers and customers. *Measurement and monitoring* of the activities of the supply chain partners are essential for successful integration. This mechanism increases the accountability of the partner firms.

5.3 Dimensions of Interoperability

Interoperability of the information system of the partner firms is a necessary condition for supply chain integration. The first requirement for interoperability is the physical level connectivity between the system through the computer networks and using communication protocols like TCP/IP. Besides physical connectivity, as per the ATHENA technical framework (Berre et al., 2007), four dimensions of interoperability are identified: data, service, process, and business.

5.3.1 Interoperability of Data

Typically the information systems of supply chain partners are built upon different technologies for operating systems and database management systems. The interoperability of data is concerned with two things: first, finding and sharing information from heterogeneous databases, and which reside on different machines with various operating systems and database management systems; second, resolving the semantics differences that exist with the structure and contents of the databases (Chen et al., 2008). Data exchange standards such as EDI and XML are primarily used to solve this level of interoperability issue.

5.3.2 Interoperability of Service

Interoperability of service is concerned with identifying, composing, and operating together various applications that are designed and implemented independently. Connecting diverse applications requires a common architectural framework that enables integration. CORBA (Common Object Request

Broker Architecture) proposed by OMG (Object Management Group) is one such example. More recent technology solutions are based on Service-Oriented Architecture (SOA). Web services is one such technology. Thus, the service interoperability deals with the capability of exchanging services (works, activities) among partners (Chen et al., 2008).

5.3.3 Interoperability of Process

A business process is defined as a set of one or more linked procedures (activities, works, and services), which collectively realize a business objective or policy goal, normally within the context of an organizational structure defining functional roles and relationships. Integration at the business process layer deals with the semantics of interactions that correspond to joint business processes. For example, the following steps constitute a joint business process: send order, process order, deliver product, and make payment. The process interoperability means linking different process descriptions to form collaborative processes. In a Web services framework such processes can be composed by putting individual services together using orchestration or choreography technologies. If the processes are orchestrated, then a single centralized service controls the process flow by calling the services in desired sequence. In case of choreography, the control is decentralized and the services are designed to call each other to maintain the sequence of operations to complete the process. Usually different process description languages are used to define process models for various purposes. Typical barriers that prevent process interoperability are different semantics and syntax used in process modeling languages, incompatibility of process execution engines and platforms, process organization mechanisms, and configurations and managements of composite processes.

5.3.4 Interoperability of Business

Business interoperability is concerned with how business processes are understood and shared without ambiguity among interoperation partners. Business interoperability explores interoperability from a business perspective and identifies the fundamental artifacts related to business issues. These issues range from the business vision and culture to the ICT infrastructure support as well as the compatibility between different organization structures, methods of work, accounting systems and rules, labor legislations, and so forth. Developing business interoperability means finding ways to harmonize those issues or at least understand them through necessary mappings and negotiations. For example, the tax calculation process while preparing the invoice may vary between both the organizations depending on the rules followed in their respective geographical regions.

5.4 A Motivating Example: Vendor-Managed Inventory Systems

Supply chain collaborations have been strongly advocated by consultants and academics under the banner of concepts such as vendor-managed inventory (VMI); collaborative planning, forecasting, and replenishment (CPFR); and continuous replenishment (CR). VMI is an efficient replenishment practice designed to enable the vendor to respond to demand without the distortive effect of the purchasing decisions (Howleg et al., 2005). VMI is widely adopted as a replenishment solution in large retail chains.

5.4.1 Integration Issues

In a VMI setting the customer (a retailer or wholesaler) gives the responsibility for placing a replenishment order to the supplier. Having the full visibility of the stock level at the retailer's site, the supplier is wholly responsible for managing the retailer's inventory. Thus, indirectly the supplier takes the responsibility of the service level of the retailer. Figures 5.1 and 5.2 represent the models of typical supplier- and retailer-side activities in a VMI setting using a Unified Modeling Language (UML) state chart diagram. It is evident that the activities at both sides are not independent of each other. Figure 5.3

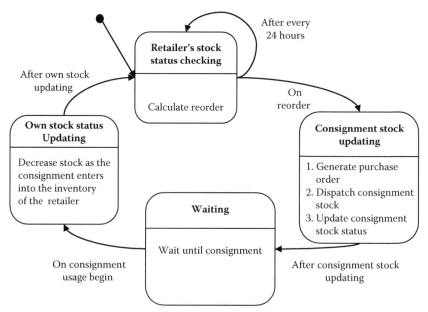

FIGURE 5.1
Modeling typical supplier-side activities in a typical vendor-managed inventory (VMI) setting using the state chart diagram.

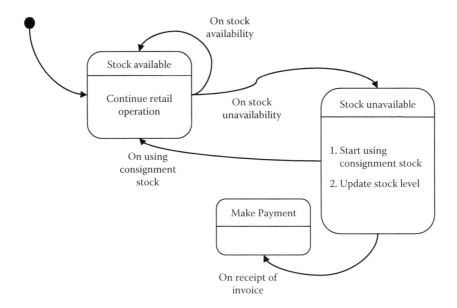

FIGURE 5.2
Modeling typical retailer-side activities in a typical vendor-managed inventory (VMI) setting using the state chart diagram.

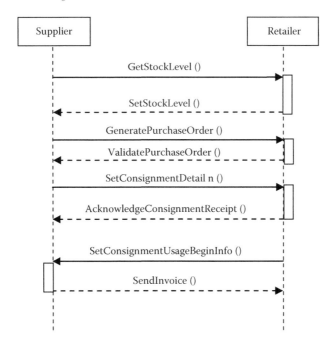

FIGURE 5.3
Modeling information flow between supplier and retailer using a sequence diagram.

is a sequence diagram showing the sequence of information flow between a retailer and a supplier in a VMI setting. As shown in the figures a supplier being responsible for continuous stock replenishment at the retailer's side checks the retailer's stock status every day, forecasts demand, and calculates the reorder point. At the reorder point, the supplier generates the purchase order for the retailer, and sends the consignment stock and the related information such as shipping details. The retailer acknowledges the receipt of the information. In a VMI setting the consignment stock, though remaining at the retailer's premise, belongs to the supplier. The retailer can defer the payment for this stock until the actual usage begins. When the retailer becomes out of stock, the retailer opens the consignment stock and starts using it. From this time onward, the retailer owns it and the retailer's stock gets updated. Therefore, as soon as the consignment stock is opened, the related information is sent to the supplier. Both supplier and retailer now adjust their stock level and the invoice is sent to the retailer. From this point onward, the supplier again continuously checks the stock status of the retailer. This cycle continues enabling the coordination between the retailer and the supplier (Singh and Jenamani, 2007).

A VMI system requires integration with the upstream supply chain partner; hence, it has a supplier focus in the horizontal dimension. The business processes in each of the information exchange dimensions for the VMI process are as follows:

Information integration: The supplier is responsible for maintaining the retailer's service level. Therefore, the supplier's information system needs to be integrated with that of the retailer to share point-of-service (POS) data and inventory status.

Planning synchronization: The supplier forecasts the demand and prepares the purchase order and sends it to the retailer. The retailer may request to modify the purchase order if certain exceptions arise, such as the fill rate is not satisfied or the retailer expects the demand to go up, or accepts it. This synchronization is essential to optimize the inventory level.

Workflow coordination: The whole VMI system can be viewed as two interrelated cycles: replenishment cycle and payment cycle. Each cycle is a workflow consisting of a set of activities linked together. Figures 5.1 and 5.2 show both cycles.

Monitoring and measurement: Monitoring of retailer and vendor activities is essential to the success of VMI processes. A VMI operation consists of three flows: information flow, material flow, and credit flow. Three flows are to be synchronized to make a successful VMI system. Suppose the retailer starts using the consignment stock but forgets to report it to the vendor (i.e., does not update the status in the VMI system), then the replenishment cycle gets disrupted.

5.4.2 Interoperability Issues

In the proposed VMI setting the supplier and retailer have their own information systems. The information system in each company may be an ERP system that connects all the functional areas of the corresponding company. Smooth operation of the proposed VMI system requires seamless integration of both the information systems. The integration has to happen at many levels for true B2B interoperability.

5.4.2.1 Interoperability of Data

The interoperability of data is concerned with two things: finding and sharing information from heterogeneous databases and which reside on different machines with different operating systems and database management systems, and resolving the semantics differences that exist with the structure and content of the databases (Chen et al., 2008). Typically the information systems for the retailer and the supplier have different technologies for operating systems and database management systems. For example, the buyer's system may be a Linux Server with a DB2 database, whereas the supplier's system is a Windows server with Oracle database. The relational data models, product coding techniques, and data security policies at both ends may also be different. For example, the same product, say a 200 gm can of fruit juice of a certain company may have the product code at JC200T1 at the retailer, whereas the same product may be called FJCan at the supplier. The attributes that define these products may also be different.

5.4.2.2 Interoperability of Service

The service interoperability deals with the capability of exchanging services (works) among partners. It is concerned with identifying, composing, and operating together various applications (designed and implemented independently) by solving the syntactic and semantic differences as well as finding the connections to the various heterogeneous databases. For example, in the VMI setting, say the ERP system at the retailer's end collects the daily sales data from the *point of sales systems* continuously. This sales data updates the inventory records. At the end of each day the inventory report must be prepared and the supplier must get access to this report. This can be possible in many different ways.

The first option can be that the report is straightway sent *through e-mail.* Then at the supplier's end some operator reads the report, manually resolves the syntactic and semantic problems with the data, and enters it into their database. The second option can be that both supplier and buyer agree upon common data syntax and semantics, possibly by adopting a *common data interchange standard like EDI.* The report file from the retailer's ERP system now gets translated using some software to the common format. The translated file may be sent through e-mail. The operator at the supplier side uses some software to retranslate the file to a format readable by the supplier's ERP system. A program at the ERP system can read the file and put the data into the database system. The third way can be to upload the inventory file in the retailer's secure *Web site*, possibly in the common standard format. The supplier's operator downloads this file at his convenience and feeds into ERP system. The last two approaches avoid re-keying the data and reduce the associated data entry errors. The fourth way of sending the data can be that the retailer hosts an XML-based service to send the inventory status data. It waits for the request by the supplier client to consume the service. The data follow a common standard format. The supplier client accesses the service, gets the XML data, and stores the data into the database without any human intervention. Such types of services are generally termed *Web services*.

5.4.2.3 Interoperability of Process

The process interoperability means linking different process descriptions to form collaborative processes. In the VMI setting, accessing the inventory status is not the only service to realize the VMI operation. As shown in Figure 5.3, for each information flow a service is to be hosted at the data generation point and a client is to be hosted where the data are consumed. All these Web services can be accessed independently to achieve service-level interoperability. But, in order to automate the complete information flows in a VMI process, these services are to be composed together by the process of *orchestration* or *choreography*. Typically barriers preventing process interoperability are different semantics and syntax used in different process modeling languages, incompatible process execution engines and platforms, different process organization mechanisms, configurations, and managements.

Seamless integration and interoperability between the retailer and the supplier enterprises are essential for complete automation of information flow for VMI system implementation. Resolving the above issues can make such automation possible.

5.5 Conceptual and Implementation Frameworks for Interoperable Systems

The domain of B2B integration and interoperability is a subdomain under the broad area of enterprise integration and interoperability. *Enterprise* is defined as one or more organizations sharing a definite mission, goals, and objectives to offer an output such as a product or a service (ISO 15704, 2000). This broad definition covers the extended enterprise, virtual enterprise, and supply chains. Enterprise integration is the process of ensuring the interaction between enterprise entities necessary to achieve domain objectives. It consists of breaking the organizational barriers to improve synergy within the enterprise so that the business goals are achieved in a more productive and efficient way. In this section we discuss two issues: first, we introduce historical project initiatives taken up to design architecture frameworks for enterprise integration and interoperability; second, we discuss recent technology solutions that have helped in realizing the proposed conceptual framework.

5.5.1 Architecture Frameworks

An architecture framework provides the conceptual modeling (i.e., enterprise modeling and implementation guidelines for integration and interoperability of large-scale systems such as a supply chain). An enterprise model is used as a semantics unification mechanism or a knowledge-mapping mechanism to achieve interoperability.

The initiatives in the 1980s for designing architecture frameworks reflect the background/competencies of their developers and the purpose for which they are elaborated: for example, CIMOSA for computer-integrated manufacturing, GRAI for production management and decision making, PERA for system engineering, Zachman for information systems, and DoDAF for military operations management and coordination. For a better understanding of these approaches and comparison between them, it is necessary to map them to a unique reference architecture. The most suitable one currently available is the ISO 15704 (Chen et al., 2008).

A few major interoperability frameworks developed after 2000 include IDEAS (2002), ATHENA Interoperability Framework (ATHENA, 2006; Berre et al., 2007), INTEROP NoE (2006), and EIF (2006). All these projects are funded under the European Commission under the Fifth and Sixth framework programs (1998–2002, 2002–2006). The IDEAS interoperability framework is the first initiative carried out to address enterprise and manufacturing interoperability issues. It is used as a basis to elaborate a roadmap served to build ATHENA Integrated Project and INTEROP NoE. The IDEAS project suggested that interoperability is achieved on multiple levels: inter-enterprise coordination, business process integration, semantic application integration, syntactical application integration, and physical integration. The major contribution of IDEA is the identification of major research domains contributing to the address interoperability issues—enterprise modeling, architecture and platform, and ontology. The INTEROP NoE project identified the barriers and concerns for interoperability. The project suggests that the barriers to interoperability can be conceptual, technological, or organizational. The conceptual barriers deal with syntactic and semantics differences of the information to be exchanged. The technological barriers refer to technological incompatibility of information and communication technologies used in the systems. The organizational barriers deal with the incompatibility of organizational structures relating the definitions of authorities and responsibilities.

More recent approaches to interoperability include Model Driven Interoperability Architecture (MDIA) and Service-Oriented Architecture (SOA) for Interoperability. MDIA clearly distinguishes between the business and computational levels. The business level is realized through enterprise and system models to create the ontologies of individual companies. These ontologies are mapped with one another to achieve syntactic and semantic harmony. Later they are used by the IT applications of individual companies to achieve interoperability. In fact both ATHENA and INTEROP NoE lay the foundations for MDIA. SOA is more technology oriented, using service-oriented paradigms such as Web services, peer-to-peer (P2P) services, and grid services.

Interoperable systems require both technical- and semantics-level interoperability. Technical interoperability is achieved if the software in both systems interacts. Semantic interoperability ensures both the systems have the same understanding of different concepts. For example, when system A refers to a concept C, system B must have the same understanding of concept C. Say, if C is a document and is called a purchase order by system A, containing certain components such as PO number, date, list of items, quantities, and item descriptions, then system B must also have the same understanding of the purchase order.

5.5.2 Implementation Frameworks

B2B interoperability can be achieved at three different levels: content level, data level, and business process level with the existing technology solutions (Dabous et al., 2003; Jones, 2001). The major difference among these levels lies in the degree of automation achieved in connecting the supply chain partners by resolving the semantic-level interoperability as discussed in the last section. Each level of integration rests on common technologies such as relational database, three-tier architecture, and internetworking protocols such as TCP/IP and FTP. Connectivity to the Internet is a must for any kind of integration.

5.5.2.1 Content Level

The content level is concerned with the exchange of messages among partners using its Web site. Under this type of integration the Web site acts as the information hub for its supply chain partners. The company posts information to be shared by the partners in the Web site. With appropriate access rights, the partners can access this information. The company may get required information from the partners through appropriate forms that establish single-sided database connectivity. HTTP is the protocol for a Web site–based communication model. For creating static pages HTML is used. A numbers of technologies such as JSP, PHP, and ASP are used to create

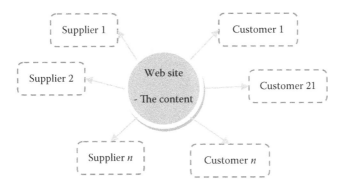

FIGURE 5.4
Content-level integration: the hub-and-spoke model.

dynamic Web pages with database connectivity. The site can be hosted in-house with connectivity to the Internet backbone through an ISP and getting a domain name, or hiring Web space from some third-party service provider. Content-level integration is the lowest cost solution toward supply chain integration. However, such a hub-and-spoke model (Figure 5.4) is not suitable for business process automation because the information provided by the site is in HTML format, which is difficult to parse automatically. Therefore, there is more human involvement in this case. Automation through data-level semantic compatibility and workflow kevel coordination cannot be achieved in this model. Therefore, the true B2B interoperability cannot be achieved at this level of integration.

5.5.2.2 Data Level

Data-level integration resolves semantic and structural heterogeneity issues. Semantic differences come from different interpretations of the same concept. For example, a data item called "price" can mean a price that includes or excludes tax. Structural differences arise from the use of diverse information formats. For example, it is difficult to automatically determine if a document represents a purchase order or a request for a quote or a product description. The interoperability objective of the data level is to provide independence from data models, formats, and languages. Proposed solutions are based on information translation and integration. Supply chain partners agree on the specific data format to encode, interpret, send, and receive data. Having nonstandard data formats (incompatible data format) would not allow different components in different systems to interact unless extra adapter software is provided to convert between the two different data formats. This difficulty becomes very clear when considering distributed systems integrated with the Web, as has been discussed in content-level integration where the Web browser acts as a client. Having a standard data format would allow new business systems not to bother about how their Web clients would interpret exchanged data. Two major data-level middleware are Electronic Data Interchange (EDI) and a number of XML-based frameworks (Figures 5.5 and 5.6).

5.5.2.3 Business Process Level

Business process level integration ensures seamless information flow connecting business processes of the supply chain partners. A business process is defined as a set of one or more linked procedures or activities that collectively realize a business objective or policy goal, normally within the context of an organizational structure defining functional roles and relationships. Integration at the business process layer deals with the semantics of interactions that correspond to joint business processes. For example, the following steps constitute a joint business process: send order, process order, deliver product, and make payment. This layer resolves issues such as what is the meaning of a message, what actions are allowed, what responses are expected, and so forth. Therefore, the layer's interoperability objective is to allow transparent peer-to-peer interactions with any partners. This is a very difficult problem. Among potential solutions are Application Programming Interface (API), workflow-based solutions, and more recently Web services.

Table 5.2 compares the three levels of integration based on seven different factors: (1) maturity level of the underlying technology; (2) use of common technology with all the business partners; (3) data-, document-, and process-level semantic compatibility with the business partners; (4) workflow compatibility achieved by seamless integration; (5) real-time access of the data generated at the source; (6) level of automation by decreasing direct human involvement; and (7) investment required by the supply chain partners to achieve interoperability. The content-level integration requires common and mature technologies such as HTTP, HTML, and so forth. Therefore, it can be used to cater to all the business partners and does not require any additional investment for integration. However, the scope of automation is limited. The

TABLE 5.2

Features of Different Levels of Interoperability

Level	Technology Maturity	Technology Compatibility	Semantic Compatibility	Workflow Compatibility	Real-Time Information Access	Level of Automation	Investment Required by the Partner
Content	High	Yes	Low	No	No	Low	No
Data	Medium	May not be	High	No	No	Medium	Yes
Process	Low	May not be	High	Yes	Yes	High	Yes

degree of human involvement is quite high for downloading the information from a Web site and re-keying the data into a database to deal with the semantic incompatibility problems. The data-level integration using EDI requires significant investment by the supply chain partners. However, the advent of Internet-based EDI using XML has significantly reduced the cost of data-level integration. EDI technology is two decades old and has been used for collaborative planning forecasting and replenishment. The use of standards in EDI solves the semantic incompatibility problems and eliminates manual data entry. The EDI files are generally sent via mail and are opened by the business partner at his or her convenience. Therefore, the workflow level of integration is not achieved. The business process–level integration shows promise for complete automation of interorganizational workflow, but the technology is evolving and requires the involvement of both supply chain partners to develop a complete solution. In the next section we discuss various technologies and standards required for supply chain integration.

5.6 Technologies for Interoperability

In this section we discuss the primary and supporting technologies and standards for achieving technical and semantic interoperability. Primary technologies facilitate information sharing between two supply chain partners. The primary technologies we discuss include EDI, XML, APIs, traditional workflow systems, and Web services. Supporting technologies fortify the primary technologies. For example, data and communication security technologies are necessary for safe electronic transactions, and RFID technologies are necessary for better inventory management. In this technology category we focus on security, payment, and RFID technologies.

5.6.1 Electronic Data Interchange (EDI)

EDI is the oldest of the integration technologies. Even with the advent of new technologies for workflow automation such as Web services, EDI is still the data format used by the vast majority of electronic commerce transactions in the world. It is a set of standards for structuring information to be electronically exchanged between the information-sharing organizations. The standards describe structures commonly used for business documents, for example a purchase order. The term *EDI* is also used to refer to the implementation and operation of systems and processes for creating, transmitting, and receiving EDI documents. The National Institute of Standards and Technology defined EDI as follows:

> The computer-to-computer interchange of strictly formatted messages that represent documents other than monetary instruments. EDI implies a sequence of messages between two parties, either of whom may serve as originator or recipient. The formatted data representing the documents may be transmitted from originator to recipient via telecommunications or physically transported on electronic storage media (http://www.patenthome.com/copyright-terms/electronic-data-interchange/).

EDIs save unnecessary recapture and re-keying of data. This leads to faster transfer of data, far fewer errors, less time wasted on exception handling, and hence a more streamlined business process. EDI also eliminates other paper-handling tasks.

The basic steps of EDI are preparation of electronic documents, outbound translation, communication, inbound translation, and processing of electronic documents. As shown in Figure 5.5, the first step in any sequence of EDI is the collection and organization of data by the sender's internal application systems. The next step is to translate this electronic file into a standard format. The sender's computer then automatically makes a connection with its Value Added Network and transmits all the files that have been prepared. The receiver retrieves the files from his or her electronic mailboxes at his or her convenience, and reverses the process that the sender went through, translating the file from the standard format into the specific format required by the receiver's application software. The receiver then processes the received document in his or her internal application systems. Examples of EDI standards are Standard of Automotive Industry Action Group (AIAG); X.12 de facto umbrella standard in the United States and Canada; EDI for Administration, Commerce, and Trade (EDIFACT) umbrella of standards in Europe. Larger companies purchase hardware, software, and proprietary networks for EDI implementation. Medium and small companies seek third-party services for value-added networking (VAN) for a fee. More recently EDI over the Internet (EDIINT) has emerged as a low-cost solution to EDI. EDI over the Internet (EDIINT) is a working group of the Internet Engineering Task Force (IETF). Example of EDIINT Standards are AS1(RFC 2821) and AS2(RFC 2616). Figure 5.5 shows a typical EDI conversion and file transfer process.

The set of supported document types is limited in EDI. This means that EDI is limited to enable a rich set of possible B2B interactions. In addition,

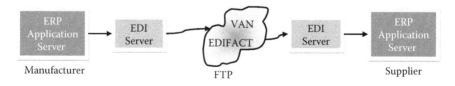

FIGURE 5.5
Data-level integration through electronic data interchange (EDI).

EDI standards, as currently defined, do not support interoperability at the business process level.

5.6.2 Extensible Markup Language (XML)

XML has created a mechanism to publish, share, and exchange data using open standards over the Internet. Every relational database management system (DBMS) has different internal formats to store the data files. Therefore, it cannot be read directly by another DBMS software. For example, as shown below the manufacturer's purchase order comes from his or her database (say DB2), which is to go into the supplier's database (say Oracle). For this process the XML wrapper, a module provided in the manufacturer's database, converts the purchase order file to XML format. This XML file will be sent over the Internet to the supplier, where an XML parser provided with this database converts it to the appropriate format (Figure 5.6). XML wrapper and parser have become integral parts of every DBMS package, making XML a common flexible data format.

XML is not, however, an integration solution in itself—it is just a data definition language. Without global XML standards there can be no seamless business among companies spread out all over the world. These standards are a common set of industry-specific definitions representing business processes. For XML messages to be interpreted by all companies participating in B2B integration, they need to agree on a common XML-based B2B standard, which defines the document, formats, allowable information, and process descriptions.

ebXML (electronic business using Extensible Markup Language) is a single set of internationally agreed upon technical specifications and common XML semantics to facilitate global trade. The ebXML framework for e-business is a joint initiative of UN/CEFACT and OASIS (www.ebxml.org). ebXML is unique in the breadth of its standards vision and is built on the previous Open-EDI standards efforts toward a shared global Internet-based B2B framework. ebXML is complementary to many existing standards, such as legacy EDI, XML-based business document standards, and Web services. Besides this common standard, the need for industry-wide B2B e-commerce standards in vertical industries is becoming increasingly critical and obvious. Several organizations have been working to define these market segment–specific definitions. Standards and groups such as RosettaNet, CIDX, and OASIS are making

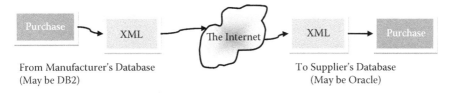

From Manufacturer's Database
(May be DB2)

To Supplier's Database
(May be Oracle)

FIGURE 5.6
Data-level integration through Extensible Markup Language (XML).

it possible for companies to share information with one another without having to completely reengineer their internal applications. These standards automate the flow of information across all companies within a given industry, independent of the underlying software or hardware infrastructure supporting the activities related to these transactions (Bussler, 2002).

5.6.3 Application Programming Interface (API)

APIs are an attempt to connect the business processes of the business partners. An API consists of working out business processes offline, determining the overall connection and coordination of operations, and then defining universally agreed upon abstract interfaces that provide remote operation invocations and connectors for back-end systems. Middleware and database technologies are then used for mapping these abstract interfaces to physical implementations. Examples of such approaches include CORBA-based solutions and Java RMI.

5.6.4 Traditional Workflow Systems

These systems are based on the premise that the success of an enterprise requires the management of business processes in their entirety. Current business processes within an organization are integrated and managed using either ERP systems such as SAP/R3, Baan, or People-Soft, or various workflow systems like IBM's MQ Series Workflow or integrated manually on an on-demand basis. However, B2B e-commerce requires the flexible support of cross-enterprises relationships. An interenterprise workflow (IEWf), is a workflow consisting of a set of activities that are implemented by different enterprises; that is, it represents a business process that crosses organizational boundaries. There are several efforts to deal with IEWfs' issues such as managing dynamic relationships among heterogeneous and autonomous organization's WfMSs and supporting scalability and availability. However, no current approach seems to provide a complete generic solution.

5.6.5 The Web Services Model

The Web services model is emerging steadily as a loosely coupled, document-based integration framework for Internet-based applications. Unlike traditional workflows, Web services support interenterprise workflows where business processes across enterprises can interact in a loosely coupled fashion by exchanging XML document-based messages. The Web services model idea is to break down Web-accessible applications into smaller services. These services are accessible through the Internet. They are self-describing and provide semantically well-defined functionality that allows users to access and perform the offered tasks. Such services can be distributed and deployed over a number of Internet-connected machines. In a typical Web services model (Figure 5.7) a *service provider* describes a service, publishes the service, and

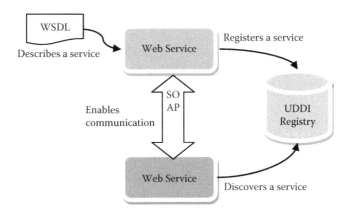

FIGURE 5.7
Web services model.

allows invocation of the service by parties wishing to do so. A *service requester* may request a service location through a *service broker* that also supports a service search. Web services are loosely coupled allowing external applications to bind to them. Web services are also reusable allowing many different parties to use and reuse a service provided (webservicesarchitect.com).

Simple Object Access Protocol (SOAP) is an XML-based protocol for exchanging document-based messages across the Internet. *Web Services Description Language* (WSDL) is a general-purpose XML-based language for describing the interface, protocol bindings, and the deployment details of Web services. *Universal Description, Discovery, and Integration* (UDDI) refers to a set of specifications related to efficiently publishing and discovering information about Web services and enables applications to find Web services, either at design time or runtime. In conclusion, the Web services model uses XML as the basis to enable SOAP at the communication layer and interenterprise workflow at the business process layer.

A typical business process is not limited to a single Web service. All the services involved in automating a business process are to be synchronized either through the process of *orchestration* or *choreography*. In the case of orchestration, a single centralized service controls the process flow by calling the services in a desired sequence. In case of choreography the control is decentralized and the services are designed to call each other to maintain the sequence of operations to complete the process. Usually process description languages such as BPEL are used to define such models.

Though the web services promise complete automation of interorganizational business processes, the underlying standards are not stabilized, creating a threat for incompatibility issues when new standards come in. Thus, if a company providing Web service interfaces for its partner adopts a new standard, the partner has to follow the same procedure to resolve the compatibility issues. It is particularly detrimental to small firms with a limited budget for IT investments. A promising technology that has the potential to solve this problem is the grid service. It is the next big venture toward solving the business process integration issues. It promises to provide a common pool of generalized services from which business process can be constructed on the fly.

5.6.6 Security Technologies

The interconnectivity of information systems in a supply chain presents a situation where small efforts can cause potentially large losses due to security lapses. Both accidental and intentional breaches are easier and more likely in the open environment like the Internet. It is a major challenge for supply chain integration. Concern for information security is fairly widespread. Banking, health care, finance, and telecommunications sectors rate information security as the highest business priority, with retailers a little less concerned. In every sector, security is regarded as a key business driver.

5.6.6.1 Security Categories

Effective information security policy must have the following six objectives: confidentiality, integrity, legitimate use (identification, authentication, and authorization), nonrepudiation, availability, and auditing or traceability (Otuteye, 2003; E-business Resource Group Security Guidelines, 2003).

Confidentiality deals with protecting the content of messages or data transmitted over the Internet from unauthorized people. For example, it is essential during information sharing in order to avoid industrial espionage. *Cryptography* tools are used to resolve the confidentiality requirements. *Integrity* is related to preventing data from being modified by an attacker. Technology involved here are *cryptography* and use of the *hash function*. *Legitimate use* has three components: identification, authentication, and authorization. Identification involves a process of a user positively identifying itself (human or machine) to the host (server) with which it wishes to conduct a transaction. The most common method for establishing identity is by means of username and password. The response to identification is authentication. Authentication usually requires the entity that presents its identity to confirm it either with something the client knows (e.g., password or PIN), something the client has (e.g., a smart card, identity card), or something the client is (biometrics: fingerprint or retinal scan). The approach to authentication that is gaining acceptance in the e-business world is the use of *digital certificates*. The authorization ensures that the authorized client accesses those parts of data or system that the client is entitled to. *Nonrepudiation* is an attribute of a secure system that prevents the sender of a message from denying having sent it. A *digital certificate* is again a solution here.

Availability ensures that system components provide continuous service by preventing failures that could result from accidents or attacks. The most commonly known cause of availability problems are host problem, network issues, and *denial of service (DoS)* attacks. In an e-business security context, *auditing* is the process of examining past transactions. Trust is enhanced if users can be assured that transactions can be traced from origin to completion. If there is a discrepancy or dispute, it will be possible to work back through each step in the process to determine where the problem occurred.

5.6.6.2 Cryptography

Cryptography is a technique by which data, called *plaintext*, is scrambled or *encrypted* in such a way that it becomes extremely difficult, expensive, and time consuming for an unauthorized person to unscramble or *decrypt* it. The encrypted text is called the *ciphertext*. The encryption and decryption are done using the security keys. Cryptographic algorithms can be classified into two broad classes: symmetric key cryptography or asymmetric key cryptography. If the keys for the encryption and decryption are the same, then the corresponding algorithm is called a *symmetric key algorithm*. Examples include DES (Data Encryption Standard), TDES, IDEA, RC2, RC4, and RC5. These algorithms can be implemented either in software form or in hardware form. The hardware implementation is typically 100 times faster than the software implementation. The major problem associated with the symmetric key algorithm is *key distribution*. Another disadvantage of this is the keys are to be securely distributed before starting the actual secure communication. A symmetric key algorithm cannot be used for authentication or nonrepudiation of the communication process.

If the encryption key is not equal to the decryption key, then the corresponding algorithm is called an *asymmetric key cryptography algorithm*. The entities wishing to use the algorithm must possess a pair of keys: a public and a private key. The public key is known to all the outside entities. RSA is an example of an asymmetric cryptography algorithm. Asymmetric key algorithms are much slower than symmetric key cryptographic algorithms. For example, RSA is 100 times slower than DES. In real-life situations RSA is never used for bulk data transfer. Rather it is used for bulk encryption key (symmetric key) exchange.

5.6.6.3 Digital Signature

A digital signature scheme is used to simulate the security properties of a signature in the digital world. It is based on the ideas of asymmetric cryptography. Digital signature schemes normally give two algorithms, one for signing which involves the user's secret or private key, and one for verifying signatures, which involves the user's public key. The output of the signature process is called the *digital signature*. Digital signatures, like written signatures, are used to provide authentication of the associated input, usually called a *message*. Messages may be anything, from electronic mail to a

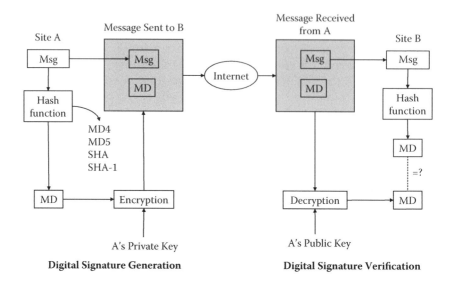

FIGURE 5.8
Digital signature generation and verification process.

contract, or even a message sent in a more complicated cryptographic protocol. The process of digital signature generation and verification is shown in Figure 5.8.

As shown in Figure 5.8, during the signature generation process, an encrypted message digest is generated and sent along with the original message. A message digest (MD) is generated by applying a hash function to the original message. A hash function is a reproducible method of turning some kind of data into a (relatively) small number that may serve as a digital fingerprint of the data. The algorithm substitutes or transposes the data to create such fingerprints. Next, the MD is encrypted using the sender's private key. At the receiver's end the MD is again recreated from the message. The encrypted MD is decrypted using the sender's public key to ensure that it is actually sent by the authentic part. The MD generated from the decryption process gets compared with the regenerated MD. If they match, it is ensured that the content of the original message is not tampered with. Thus, a digital signature ensures data integrity and nonrepudiation.

Digital signatures are used to create public key infrastructure (PKI) schemes in which a user's public key (whether for public-key encryption, digital signatures, or any other purpose) is tied to a user by a digital identity certificate issued by a certificate authority. PKI schemes attempt to unbreakably bind user information (name, address, phone number, etc.) to a public key, so that public keys can be used as a form of identification.

5.6.6.4 *Public Key Infrastructure and Digital (Identity) Certificate Generation*

Public key cryptography supports security mechanisms such as confidentiality, integrity, authentication, and nonrepudiation. To successfully implement these security mechanisms, an infrastructure to manage them should be planned. A public key infrastructure (PKI) is a foundation on which other applications, systems, and network security components are built.

The framework of a PKI consists of security and operational policies, security services, and interoperability protocols supporting the use of public-key cryptography for the management of keys and certificates. The generation, distribution, and management of public keys and associated certificates normally occur through the use of Certification Authorities (CAs), Registration Authorities (RAs), and directory services, which can be used to establish a hierarchy or chain of trust. CA, RA, and directory services allow for the implementation of digital certificates that can be used to identify different entities. The purpose of a PKI framework is to enable and support the secured exchange of data, credentials, and value (such as monetary instruments) in various environments that are typically insecure, such as the Internet.

The basic functions of PKI are the generation, distribution, administration, and control of cryptographic keys. PKI binds a public key to an individual, organization, or other entity, or to some other data, for example, an e-mail or purchase order. PKI is responsible for digital certificate validation. It verifies that a certificate is still valid for specific operations and ensures that a trusted relationship or binding exists. If necessary, a CA can also cancel (revoke) a previously issued certificate.

To participate in a PKI, an end entity must enroll or register in a PKI. The result of this process is the generation of a public-key certificate. The binding is declared when a trusted CA digitally signs the public-key certificate with its private key. To generate a certificate, the CA performs the following steps:

1. Acquire a public key from the end entity.
2. Verify the identity of the end entity.
3. Determine the attributes needed for this certificate, if any.
4. Format the certificate.
5. Digitally sign the certificate data.

5.6.7 Payment Technologies

A payment system is a mechanism that facilitates transfer of value between a payer and a beneficiary by which the payer discharges the payment obligations to the beneficiary. A payment system enables two-way flow of payments in exchange of goods and services in the economy. An electronic payment system helps to automate sales activities, extends the potential number of customers, and may reduce the amount of paperwork. The technologies for a payment system can be broadly classified into B2B payment systems and B2C payment systems. Three notable e-payment technologies for B2B transactions are Automated Clearing House (ACH)–based bank proprietary e-payment platforms; enhanced purchasing card technology; and open network systems. B2C payment services can be digital cash–based, cheque-based, or credit card–based payment. Out of these mechanisms, credit card–based payment is the most popular. In the context of supply chain integration, most of the transactions are B2B types (Asokan et al., 1997; Cotteleer et al., 2007).

ACH-based transactions are the oldest form of B2B payment transactions. ACH is a nationwide electronic funds transfer (EFT) system that provides for the interbank clearing of credit and debit transactions and for the exchange of information among participating financial institutions. Prior to the transaction the bank stores complete trade information such as contract, pricing, orders, receipts, and so forth, for buyer and supplier. In a typical payment transaction the payer is prompted to authorize electronic debit by entering bank routing number and account number through the Web site of a trusted third party (TTP). TTP forwards the bank routing number and account number to the payment processor. The routing number and account number are validated, and the integrity of the account's checking history is verified. The processor forwards approve/decline results to TTP. TTP returns approval/decline message to the payer. If approved, TTP routes check for settlement through a processor to the ACH. Funds are deposited in approximately 1 to 3 business days.

Purchasing cards (p-cards) are similar to credit cards. They are usually issued to employees who are required to operate within a set of company rules and guidelines that usually include an approved spending limit. P-card transactions settle through an online "credit gateway" associated with the p-card issuer which allows institutions to exchange encrypted information for authorization purposes. These systems provide a secure environment for payments because sensitive account information is not exchanged directly, and transactions are easily traceable and challenged when disputes arise. Because the transaction fee is quite high for such cards, they are seldom used in case of medium- to high-value transactions.

Open network systems, like Visa's "Visa Commerce" platform interface directly with front-end procurement and accounts-payable systems of the participant. Open networks also enable buyers to initiate and settle payments based on preestablished terms with suppliers. Unlike enhanced p-card and ACH-based platforms, open networks employ a flexible fee schedule, allowing for the minimization of fees for relatively high-dollar-amount transactions. In addition, open-network solutions may be easier to integrate with legacy environments.

5.6.8 Radio Frequency Identification (RFID) Technology

Automated identification and data capture (AIDC) includes technology to identify objects and automatically collects data about them and updates the

TABLE 5.3

Bar Code vis-à-vis Radio Frequency Identification (RFID) Solution for Automated Data Capture

Bar Code Deficiency	RFID Improved Solution
Line-of-sight technology	Able to scan and read from different angles and through certain materials
Unable to withstand harsh conditions (dust, corrosive), must be clean and not deformed	Able to function in much harsher conditions
No potential for further technology advancement	Technology advancement is possible due to new chip and packaging technique
Can only identify the items generally and not as unique objects	EPC code will enable to identify up to 2^{96} items uniquely
Poor tracking technology, labor intensive, and slow	Potential to track the items in real time as they move through the supply chain

data into software systems without human intervention. Modern AICS heavily rely upon bar codes for automated data capture. However, there are some inherent issues with using a bar code as shown in Table 5.3. To overcome these issues the industry is now adopting new-generation AIDC technology like the RFID.

Unlike bar codes, RFID tags attached to an object can remotely identify an object, even when the object is inside a container. Therefore, it is easily identified while it moves through the supply chain. Following are some of the applications of RFID technology in the supply chain:

Advanced Shipping Notice (ASN): RFID automatically detects when either a pallet or shipment has left the warehouse or distribution center. This allows a business to not only generate an electronic ASN and notify the recipient, but also to bill clients in real time instead of waiting until the end of the week or month and doing a batch operation.

Shrinkage: One of the major problems in the supply chain is product loss or shrinkage, which can account for 2% to 5% of stock. The causes may vary from misplaced orders, employee and customer theft, or inefficient stock management. RFID with its superior tracking and identification capability is able to localize where losses are occurring.

Returned goods: Tracking the returned goods is a major problem in the reverse logistics networks. Using RFID full visibility and automation can be achieved on returned goods, thereby reducing fraud.

Anti-counterfeit: Illegal duplication and manufacture of high-value products is one of the industry's most well-known problems. By integrating a tag into the item, the tag has the potential to authenticate a product and combat the sale of false goods on the black market.

Supply chain efficiency: RFID enables traceability and reduction in the number of discrepancies between what a supplier invoiced and what a customer actually received.

Improved stock management: Managing stock is a key priority for many retailers. Implementing RFID at the item level and on shelves gives an automatic way of knowing and managing stock levels.

Reduction in labor costs: At distribution centers, labor accounts for nearly 70% of costs. It is estimated that RFID could reduce this by nearly 30% by removing the need for manual intervention and use of bar codes when loading cases or stocking pallets.

5.6.8.1 Technology Overview of RFID Systems

An RFID system (Figure 5.9) is composed of three main elements: an RFID tag (inlay), which contains data that uniquely identifies an object; an RFID reader, which writes this unique data on the tags and, when requested, can read this unique identifier; and an RFID middleware, which processes the data acquired from the reader and then updates the data to the backend database or ERP systems (Potdar et al., 2007; www.rfidgazette.org).

An *RFID tag* is a microchip attached with an antenna to a product that needs to be tracked. The tag picks up signals from the reader and reflects back the information to the reader. The tag usually contains a unique serial number, which may represent information, such as a customer's name, address, and so forth. The tags can be classified based on their ability to perform radio communication: active, semiactive (semipassive), and passive tags. They can also be classified based upon their memory: read-only, read-write or write-once, and read-many. *Active tags* have a battery that provides necessary energy to the microchip for transmitting a radio signal to the reader. *Semiactive tags* also contain a battery that is used to run the circuitry on the microchip; however, it still relies on the reader's magnetic field to transmit the radio signal (i.e., information). *Passive tags* completely rely

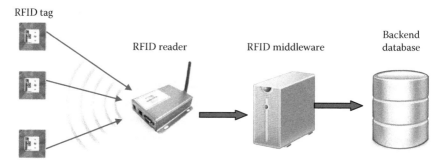

FIGURE 5.9
A typical radio frequency identification (RFID) setup.

on the energy provided by the reader's magnetic field to transmit the radio signal to and from the reader. It does not have a battery. As a result, the read range varies depending upon the reader used, the distance from which they can be accessed is much less than the other two.

RFID readers send radio waves to the RFID tags to inquire about their data contents. The tags then respond by sending back the requested data. The readers may have some processing and storage capabilities. The reader is linked via the RFID middleware with the backend database to do any other computationally intensive data processing. RFID readers can be classified using two different schemes. First, the readers can be classified based on their location as handheld readers and fixed readers. Second, the tags can be classified based upon the frequency in which they operate: single frequency and multifrequency.

RFID middleware manages the readers and extracts Electronic Product Code (EPC) (explained in the next section) data from the readers; performs tag data filtering, aggregating, and counting; and sends the data to the backend database, or enterprise WMSs (warehouse management systems) or information exchange broker. An RFID middleware works within the organization, moving information (i.e., EPC data) from the RFID tag to the integration point of high-level supply-chain management systems through a series of data-related services.

5.6.8.2 The Electronic Product Code

An EPC is designed to uniquely identify a product in a supply chain. The idea of EPC originated in the Auto-ID center in the Department of Mechanical Engineering at MIT. Later, the technology was transferred to EPC global (www.epcglobalinc.org), which administers and develops EPC standards. The EPC code is similar to the UPC (Universal Product Code) used in bar codes, and ranges from 64 bits to 256 bits with four distinct fields. For example, a 96 bit code consists of the following fields:

- Header (0–7 bits): The header defines the length of the code.
- EPC Manager (8–35 bits): Contains the manufacturer ID of the product to which the EPC tag is attached.
- Object Class (36–59 bits): This refers to the exact type of product.
- Serial Number (60–96 bits): This provides a unique identifier for up to 2^{96} products.

EPC infrastructure allows immediate access to product information from anywhere within the supply chain. It consists of three distinct components as shown in Figure 5.10: Middleware/savant, Object Name Service (ONS) database, and a Product Markup Language (PML) server. *RFID savant* is a software buffer in between the RFID readers and the servers storing the product

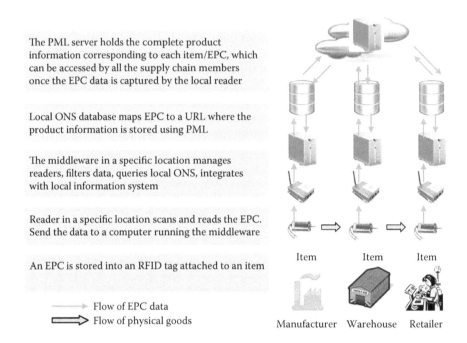

The PML server holds the complete product information corresponding to each item/EPC, which can be accessed by all the supply chain members once the EPC data is captured by the local reader

Local ONS database maps EPC to a URL where the product information is stored using PML

The middleware in a specific location manages readers, filters data, queries local ONS, integrates with local information system

Reader in a specific location scans and reads the EPC. Send the data to a computer running the middleware

An EPC is stored into an RFID tag attached to an item

Item   Item   Item

———▸ Flow of EPC data
⇒ Flow of physical goods

Manufacturer   Warehouse   Retailer

FIGURE 5.10 (See color insert.)
Using Electronic Product Code (EPC) infrastructure in the supply chain.

information. It allows companies to process relatively unstructured tag data taken from many RFID readers and direct it to the appropriate information systems. Savants are able to perform many different operations, such as monitor the RFID reader devices, manage false reads, cache data, and query an ONS. ONS matches the EPC code to information about the product via a querying mechanism similar to the DNS (Domain Naming System) used in the Internet, which is already proven technology capable of handling the volume of data expected in an EPC RFID system. The ONS server provides the IP address of a PML Server that stores information relevant to the EPC. The product information is written in a new standard software language called Physical Markup Language. PML is based on the widely used and accepted extensible markup language (XML), designed as a document format to exchange data across the Internet. The data stored in the PML server hold the key to product data accessibility throughout the supply chain. As shown in Figure 5.10, an RFID tag attached with an item contains the EPC code. The details of the item containing the code are stored in a common PML server that is accessible to all supply chain members through the EPC infrastructure. To start with, at the manufacturer the tag is written with EPC and the details are stored in the common PML server. The cases and pallets can also be identified in a similar manner. The tags now broadcast their individual EPC and can be uniquely identified. At the warehouse and the retailer when the items are unloaded, if an antenna is installed the packages need not be opened to check the details. The radio signal emitted from all the items as well as the pallet tracked by the antennas are filtered and processed by the middleware. The middleware connects to the local ONS server, which may in turn contact the PML server to get the product details of the tagged items.

Use of RFID can reduce the out-of-stocks, labor costs, inventory inaccuracies, and simplify business processes. Wal-Mart and the U.S. Department of Defense enforce vendor placement of RFID tags on all shipments to improve supply chain management. Due to the size of these two organizations, their RFID mandates impact thousands of companies worldwide. The need to organize and make decisions based on the data provided by the RFID tags is prominent. To date, research that conjoins RFID technology and item-level inventory management on the shop floor is at a preliminary stage, only inferring benefits upon application. There are many challenges to RFID implementation that include making the layout for optimal placement of antennas, collecting RFID data in a timely manner, processing such voluminous data, taking care of security and privacy issues, and making timely decisions.

5.7 Object-Oriented Analysis for Integrated Supply Chain Design and Implementation

Implementation of an integrated supply chain calls for analysis of inter- and intraorganizational business processes. A business process is described as a procedure relevant for adding value to an organization. It is viewed in its entirety, from beginning to end (Scheer and Nüttgens, 2000). A business process in the context of a supply chain involves entities from partnering organizations. The heart of the business process automation is the associated workflow. Workflow is characterized by a list of activities that need to be done to accomplish a task. There is a difference between workflow and process descriptions: workflow describes the "how" while business process is more akin to "what" and "why" (Eskeli, 2009). Object-oriented analysis has become a default tool to understand and reengineer the business process and define the underlying workflow. In this section we discuss the object-oriented analysis tools for understanding business processes. We show the usage of Unified Modeling Language (UML) tools to analyze a supply chain. We once again use examples from the VMI system setting (Section 5.3).

5.7.1 Scenario Models

A scenario is a description that illustrates, step by step, how a user is intending to use a system, essentially capturing the system behavior from the

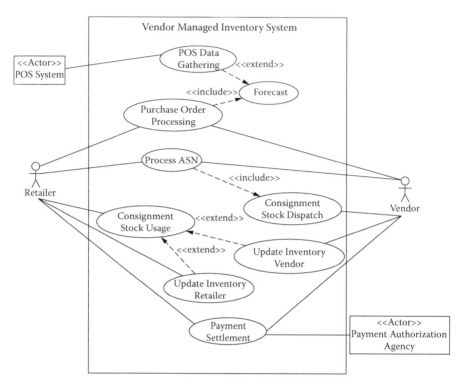

FIGURE 5.11
Use case diagram for vendor-managed inventory (VMI) system.

user's point of view. A scenario can include stories, examples, and draw-
ings. In UML, case diagrams are used as a tool to build scenario models.
Use cases are extremely useful for describing the problem domain in unam-
biguous terms and for communicating with the potential users of a system.
Figure 5.11 shows the use case diagram of a VMI system. As discussed in
Section 5.3 the diagram shows all the usage scenarios in VMI system starting from data gathering to the payment settlement process. Two primary
actors in the system are the retailer and the vendor. An actor is an entity that
participates in a use case scenario. Besides the retailer and the vendor, two
other software systems (Point of Sales System and Payment Authorization
Agency) are also actors. Each scenario appears as an oval in the figure. Many
use cases are complex and include secondary use cases. For example, a purchase order processing use case requires demand forecasting and generation
of reorder points. Similarly, forecasting is an *extension* to POS data collection.
Each use case is associated with a step-by-step description of the basic workflow involving the participating actors. The exceptions that may happen to
the basic workflow are captured in the form of alternate flow.

5.7.2 Domain Model

A domain model is a conceptual representation of a problem domain. It relates various entities and the concepts under the domain of discourse. It also has the constraints that govern the integrity of the model elements. A domain model can be conveniently represented by a UML class diagram. A class diagram consists of classes (the rectangular boxes) and the relationships (interconnecting lines). For example, consider a partial domain model of the VMI system (Figure 5.12). In this diagram we do not represent all the entities and concepts in the problem domain. Therefore, we call it a *partial domain model*. The diagram has nine different entities representing classes.

In the above model, the classes are related in the following ways:

1. The vendor creates purchase order and shipping note.
2. The vendor sends consignment stock.
3. The retailer receives and verifies the purchase order.
4. Each purchase consists of one or more items.
5. The retailer also receives a shipping note and consignment stock.
6. Each consignment is associated with a shipping note.
7. Each shipping note is associated with only one purchase order.
8. The shipping order has a source from where the consignment is picked up, a destination to deliver to, and a carrier responsible for transportation.
9. Source, destination, and carrier are specialization of address.

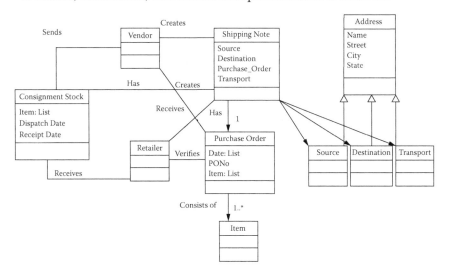

FIGURE 5.12
A partial domain model for a vendor-managed inventory (VMI) system.

5.7.3 Workflow Model

A workflow connects the activities that make one business process. Workflow can be modeled using UML activity diagrams. Figure 5.13 shows the UML activity diagram for a typical VMI process. This diagram identifies information flows and shared data objects among the actors involved in a business process. The vertical partitions called the *swimlanes* show the actors retailer and supplier participating in the VMI business process. Object flows that cross swimlanes are information flows, and they carry shared data objects. In Figure 5.13, the six information flows are denoted by numbers in the sequence in which they occur (Liu et al., 2006). The information flows are as follows:

1. The customer shares actual sales data (Point of Sales Data) with the vendor.
2. The vendor generates demand forecast and places replenishment orders for customers.

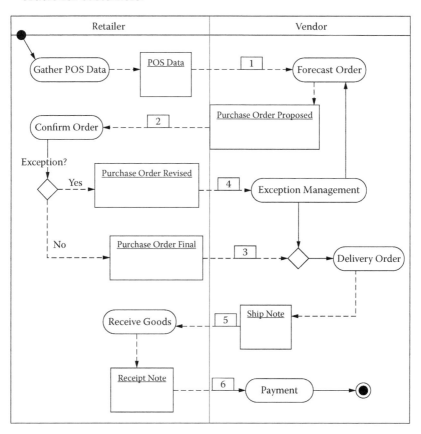

FIGURE 5.13
Workflow within the VMI system (Liu et al., 2006).

3. Customers review replenishment orders and confirm them.

4. Occasionally, there may be a need for exception handling. For example, the retailer expects the demand to go up because of some event. Under such circumstances the purchased order proposed is modified and sent by the vendor.

5. The vendor then sends ship notices, followed by physical goods transfer.

6. Customers acknowledge the actual receipt.

5.7.4 Interaction among Entities

While the class diagram shows the entities and their relationships, the sequence diagram shows the interaction among the entities. In other words, while the class diagrams show a static view of the system, the interaction diagrams show the dynamic view indicating message flow among the entities. Interactions among the entities can be modeled using either sequence or collaboration diagrams. A sequence diagram shows object interactions arranged in time sequence, while the collaboration diagrams represent the same idea by sequentially numbering the message flows. Sequence diagrams are typically associated with use case realizations in the logical view of the system under development. Each activity represented in an activity diagram can be implemented through one or more sequence diagrams. For example, the payment activity is initiated by the retailer, and the activity is closed after the vendor receives the payment. The activity can be performed by electronic funds transfer (EFT) process. Figure 5.14 shows the process of making EFT. During the EFT process the retailer entity creates a banking transaction through a payment gateway. The payment gateway is interfaced with the banking network to verify sender and receiver identities and is responsible for transferring the specified amount from the retailer's account to the vendor's account. Various operations that may be performed by different entities are shown in the diagram.

5.8 Application of Integration Technology in Supply Chain—An Example

Web service is a promising technology with the potential of providing a low-cost solution for interoperability. We discuss a laboratory-scale application of this technology to realize the information flow in the VMI system. As discussed in Section 5.3, we consider a single echelon supply chain consisting of a single supplier and a single retailer. Referring to Figure 5.3 we find a number of information flows between the supplier and the retailer.

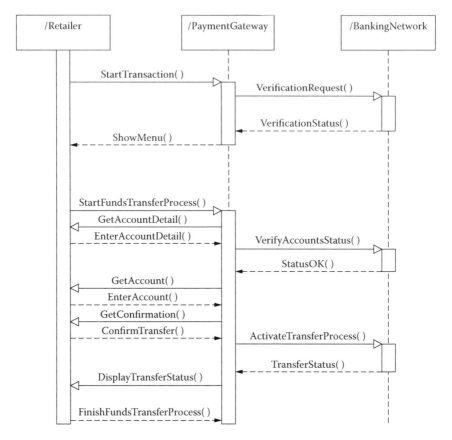

FIGURE 5.14
Sequence diagram for the payment process.

A service needs to be hosted at the source of the information (retailer/supplier), and a client is to be designed in the destination. Therefore, depending on the information flow, a number of services and clients are to be hosted and synchronized on the retailer's as well as the supplier's side. To solve this problem we design a framework for implementing VMI using composite Web services.

5.8.1 Simulation Models to Create a Demand and Supply Environment

Because we carry out this exercise at a laboratory scale, we first make two simulation models to create the demand at the retailer's end and supply at the vendor's end. The retailer's model creates random demand following the distribution defined in the simulation model. Similarly at the supplier side production is made in a simulation environment depending on the demand. Figure 5.15 shows the complete structure of database tables

Supplier Side

S1: Stock Information Table

RetailerID	ProductID	StockLevel
RT1002	P203	90
RT1002	P402	20
RT1002	P321	47

S2: Consignment Stock Table

RetailerID	CSID	ProductID	Quantity	Status
RT1002	CS201	P203	200	Used
RT1002	CS202	P402	125	Using
RT1002	CS203	P321	55	Unused
RT1006	CS402	P213	170	Using

S3: Invoice Table

InvoiceID	CSID	Amount
Inv101	CS201	20000
Inv201	CS202	30000
Inv105	CS203	100000

S4: Reorder Point Table

RetailerID	ProductID	ReorderLevel
RT1002	P203	20
RT1002	P402	20
RT1002	P321	10

Retailer Side

R1: Stock Information Table

ProdID	StockLevel
P203	90
P402	20
P321	47

R2: Consignment Stock Table

CSID	ProdID	Quantity	Status
CS201	P203	200	Used
CS202	P402	125	Using
CS203	P321	55	Unused

R3: Invoice Table

InvoiceID	CSID	Amount	Status
Inv101	CS201	20000	Paid
Inv201	CS202	30000	Paid
Inv105	CS203	100000	Pending

FIGURE 5.15
Database tables used at the supplier and the retailer sides.

required by both sides. For simulating customers at the retailer side we decrease the *StockLevel* of Table R1 depending on the demand. This information is made available to the supplier through a Web service, though with the supplier can access the information by using its client. The supplier updates the Table S1 at a regular interval and when *StockLevel* becomes less than or equal to *ReorderLevel* (shown in Table S4), the supplier sends the consignment stock to the retailer and updates the information in consignment stock Table S2. When the retailer obtains this information he or she stores it in consignment stock Table R2, and when *StockLevel* in Table R1 becomes zero, the retailer starts using the new consignment stock and changes the status in Table R2 from *unused* to *using*. This status is made available to the supplier through a Web service. The supplier updates the status in Table S2. Every time the status changes from unused to using, the vendor generates an invoice that is received by the retailer and stored in invoice Table R3. Upon payment by the retailer, an acknowledgement is received by the supplier.

5.8.2 Technology Used

We used two different programming platforms to simulate the actual conditions and also achieve the interoperability. Two widely adopted heterogeneous platforms are used: .Net and J2EE. The implementation language at the retailer side is C# (.Net) and at the supplier side it is Java (J2EE). The services are hosted in the IIS and Sun Application server, respectively. Web Services Description Language (WSDL) is used to describe the services, and the SOAP protocol is used for service implementation. The corresponding information flow is shown in Figure 5.16.

Composite Web services are implemented using Business Process Execution Language (BPEL). BPEL builds on the foundation of XML and Web services; it uses an XML-based language that supports the Web services technology stack, including SOAP, WSDL, and UDDI. BPEL is mainly used for orchestrations and choreography among Web services. For creating a composite Web

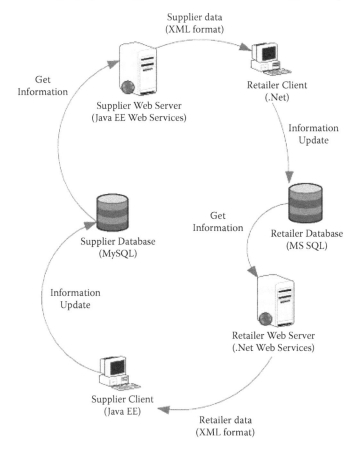

FIGURE 5.16
Data exchange between two different technologies.

service we need a BPEL engine. A BPEL engine handles Web services by three methods <receive>, <reply>, and <invoke>.

5.8.3 Design of the Composite Web Services

In the proposed framework two composite Web services are hosted in the supplier's site (Figure 5.17). The first one called CWS0 represents the

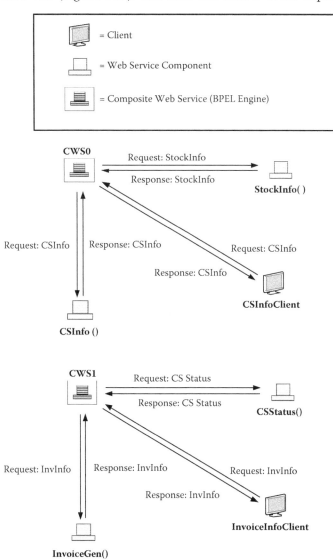

FIGURE 5.17
Composite Web services for vendor-managed inventory.

preconsignment stock delivery business process. The second one called CWS1 represents the *postconsignment stock delivery* business process.

CWS0 is initiated by a client (CSInfoClient) at the retailer's side which checks for the consignment stock status. Before sending the response to the client, CWS0 checks the retailer's stock status by coordinating two services. The one hosted at the retailer's site (StockInfo) provides the current stock update. If the stock is less than or equal to the reorder point then another service hosted at the supplier's side (CSInfo) is invoked and provides the consignment stock status. In case the status is not available, a null value is returned to the retailer client. The client continuously checks consignment stock status until it gets the non-null value (consignment stock detail). This ends the preconsignment stock delivery business process.

CWS1 is initiated by another client hosted at the retailer's site (InvoiceInfoClient). Before sending the invoice, CWS1 checks for consignment stock usage status invoking a service (CSStatus) hosted at the retailer's side. In case the consignment has been in use, CWS1 requests invoice information be sent to a service (InvoiceGen) hosted at the supplier's site. Upon getting the invoice detail it sends it back to the invoice-requesting client at the retailer's side. If the invoice is not ready, the client gets a null value and continues to check the invoice status until a non-null value is obtained. The postconsignment stock delivery business process stops here, and the preconsignment stock delivery business process starts once again.

5.8.4 Information Flow

Both clients continuously check for the changes occurring at their respective Web services. When client CSInfoClient requests for the consignment stock information from composite Web service CWS0, CWS0 invokes StockInfo() Web service to get the present stock information. If stock level becomes equal (or less) to reorder point, composite Web service CWS0 invokes CSInfo() Web service and gets the consignment information and sends it back to the CSInfoClient.

In a similar manner InvoiceInfoClient sends a request to CWS1 composite Web service to get the invoice information. CWS1 invokes the Web service CSStatus() (which keeps track of the consignment stock usage begin information), and if the present quantity becomes equal to the consignment stock quantity, then it invokes the InvoiceGen() Web service (which generates the invoice) and gets the invoice information and sends the data to InvoiceInfoClient. After getting invoice information, the retailer updates the invoice acknowledgement database. Client InvAckClient receives invoice acknowledgement by sending a request and receiving a response from Web service InvAck(). Figures 5.18 and 5.19 show the BPEL code for both composite Web services.

VMI is an effective information-sharing mechanism that can enhance the performance of the supply chain. Because the proposed implementations are

```
<process name="Consignment Information">

<partnerLinks>

        <partnerLink name="StockInfo" partnetLinkType="tns:csinfo" myRole="Sender" partner
        Role="Receiver"/>

        <partnerLink name="CSInfo" partnetLinkType="tns:csinfo" myRole="Sender" partner
        Role="Receiver"/>

        <partnerLink name="CWS0" partnetLinkType="tns:csinfo" myRole=" Receiver" partner
        Role=" Sender"/>

</partnerLinks>

<variables>

        <variable name="csinforequest" messageType="tns:csinforequest/>

        <variable name="stockinforequest" messageType="tns:stockinforequest"/>

        <variable name="stockinfo" messageType="tns:stockinfo"/>

        <variable name="csinfo" messageType="tns:csinfo"/>
```

FIGURE 5.18
Business Process Execution Language (BPEL) code for composite Web service CWS0.

```
<process name="Invoice Information">

<partnerLinks>

        <partnerLink name="CSStatus" partnetLinkType="tns:invoicegen" myRole="Sender"
        partner Role="Receiver"/>

        <partnerLink name="InvoiceGen" partnetLinkType="tns:invoicegen" myRole="Sender"
        partner Role="Receiver"/>

        <partnerLink name="CWS1" partnetLinkType="tns:invoicegen" myRole=" Receiver"
        partner Role=" Sender"/>

</partnerLinks>

<variables>

        <variable name="invoiceinforequest" messageType="tns:csinforequest />

        <variable name="csstatusrequest" messageType="tns:csstatusrequest"/>

        <variable name="csstatusinfo" messageType="tns:csstatusinfo"/>

        <variable name="invoiceinforequest" messageType="tns:invoiceinforequest"/>
```

FIGURE 5.19
Business Process Execution Language (BPEL) code for composite Web service CWS1.

costly, SMEs have not been attracted to adopt the concept. Web services can provide cost-effective ways for B2B enterprise integration and help in implementing VMI. This example shows how the information flow can be automated for VMI under a single retailer and single supplier case. The present scenario can be extended to multiple retailer or multiple supplier cases. Under such circumstances, algorithms are to be developed to schedule and sequence various Web services. The security and product code compatibility issues must also be resolved to make this model applicable to real-life situations.

5.9 Conclusions

A responsive supply chain is defined as a network of firms capable of creating wealth to their stakeholders in a competitive environment by reacting quickly and cost effectively to changing market requirements (Gunasekaran et al., 2008). IT integration of the supply chain partners holds the key to responsiveness. However, IT integration alone cannot ensure supply chain responsiveness and performance as revealed from various case studies (Themistocleous et al., 2004; Wang and Chan, 2010) and opinion surveys (Li, 2009; Wong et al., 2011). Coordination at the strategic and tactical levels is essential for true integration. In other words, if a trusted relationship exists between the business partners, then the IT integration reaps maximum benefit (Sinkovics et al., 2011). Mathematical models for designing responsive supply chains often assume that the time delay for information transfer between supply chain members is zero if their systems are integrated (You and Grossmann, 2008). In reality this is not quite true with IT systems where human intermediation is necessary. Therefore, the models for responsiveness must be designed taking care of the delays that may result due to lack of cooperation at strategic and tactical levels in spite of IT integration.

The benefit of supply chain integration is no more a matter of debate. Many theoretical models and cases emphasize these issues. However, adoption of such practices is slower than expected especially for SMEs. A study by Harland et al. (2007) reveals that while larger downstream firms are keen to adopt integration through e-business practices, the smaller upstream firms are cautious about investing in integration technology. Such firms are sometimes enforced by their downstream partners to adopt integration technologies, but they do not appreciate the full benefits to be gained due to certain barriers and mind-sets. Jharkharia and Shankar (2005) identify 10 such barriers to the IT adoption supply chain, such as threats of information security, lack of trust in supply chain linkage, and fear of information system breakdown. Other barriers to integration include technology, organizational focus, trust, people, and internal structure (Frohlich, 2002). Cheaper and open-source solutions to integration through Web technologies have a lot of promise to make the supply chain integration a reality (Ranganathan et al., 2004). But, as demonstrated by Bagchi and Skjoett-Larsen (2002), the arrival of new and cheaper technologies is not the only solution to removing the barriers without significant organizational and cultural changes.

References

Asokan, N., Janson, P.A., Steiner, M., and Waidner, M. 1997. The state of the art in electronic payment systems. *IEEE Computer,* 30(9), 28–35.

ATHENA. 2006. http://www.modelbased.net/aif/ (accessed on June 13, 2011).

Bagchi, P.K., and Skjoett-Larsen, T. 2005. Supply chain integration: A European survey. *The International Journal of Logistics Management*, 16(2), 275–294.

Berre, A., Elvesaeter, B., Figay, N., Guglielmina, C., Johnsen, S., Karlsen, D., Knothe, T. and Lippe, S. 2007. The ATHENA Interoperability Framework, Enterprise Interoperability II, Part VI, 569–580.

Bussler, C. 2002. The role of B2B engines in B2B integration architectures. *SIGMOID Record*, 31(1).

Chen, D., Doumeingts, G., and Vernadat, F. 2008. Architectures for enterprise integration and interoperability: Past, present and future. *Computers in Industry*, 59(7), 647–659.

Cotteleer, Mark J., Cotteleer, Christopher A., and Prochnow, Andrew W. 2007. Cutting checks: Challenges and choices for the adoption of B2B electronic payments. *Communications of the ACM*, 50(6), 56–61.

Dabous, F.T., Rabhi, F.A., and Ray, P.K. 2003. Middleware technologies for B2B integration. *IEC Technical Review, IEC Press*, 1–13.

E-business Resource Group Security Guidelines. 2003. http://www.bc.pitt.edu/ebusiness/arEBSecurityGuide.pdf (accessed November 20, 2011).

e-Business W@tch. 2005. Special report on e-Business Interoperability and standards, European Commission, Enterprise and Industry Directorate General.

EIF. 2006. http://ec.europa.eu/idabc/en/document/2319/5644.html (accessed June 13, 2011).

Eskeli, J. 2009. Integrated tool support for hardware related software development. Master's thesis, University of Oulu, Department of Electrical and Information Engineering, Oulu, Finland.

Frohlich, M.T. 2002. e-Integration in the supply chain: Barriers and performance. *Decision Sciences*, 33, 537–556.

Gunasekaran, A., Lai, K., and Cheng, T.C.E. 2008. Responsive supply chain: A competitive strategy in a networked economy. *Omega*, 36(4), 549–564.

Harland, C.M., Caldwell, N.D., Powell, P., and Zheng, J. 2007. Barriers to supply chain information integration: SMEs adrift of eLands. *Journal of Operations Management*, 25(6), 1234–1254.

Howleg, M., Disney, S., Holmström, J., and Smaros, J. 2005. Supply chain collaboration: Making sense of supply chain continuum. *European Management Journal*, 23(2), 170–181.

IDEAS. 2002. http://cordis.europa.eu/fetch?CALLER=PROJ_ICT_TEMP&ACTION =D&CAT=PROJ& RCN=63037 (accessed June 13, 2011).

IEEE. 1990. IEEE Standard Computer Dictionary: A Compilation of IEE Standard Computer Glossaries. (http://www.ieeexplore.ieee.org/stamp.jsp?tp=&arnumber=159342)

INTEROP NoE. 2006. http://www.interop-vlab.eu/ (accessed June 13, 2011).

ISO 15704. 2000. Industrial Automation Systems—Requirements for Enterprise-Reference Architectures and Methodologies.

Jharkharia, S., and Shankar, R. 2005. IT-enablement of supply chains: Understanding the barriers. *Journal of Enterprise Information Management*, 18(1), 11–27.

Jones, R. 2001. B2B integration. *Manufacturing Engineer*, 80(4), 165–168.

Kannan, V.R., Tan, K.C., and Narasimhan, R. 2007. The impact of operations capability on firm performance. *International Journal of Production Research*, 45(2), 5135–5156.

Lee, H. 2000. Creating value through supply chain integration. *Supply Chain Management Review*, 4(4), 30–36.

Lee, H.L., and Whang, S. 2001. E-business and supply chain integration. Stanford Global Supply Chain Management Forum. SGSCMF-W2-2001, November.

Li, G., Yang, H., Sun, L., and Sohal, A.S. 2009. The impact of IT implementation on supply chain integration and performance. *International Journal of Production Economics*, 120(1), 125–138.

Liu, R., Kumar, A., and Stenger, A. 2006. Simulation results for supply chain configurations based on information sharing. In *Proceedings of the 2006 Winter Simulation Conference*, pp. 627–635.

Otuteye, Eben. 2003. A systematic approach to e-business security, http://ausweb. scu.edu.au/ aw03/papers/otuteye/paper.html (accessed Novemeber 20, 2011).

Potdar, V., Wu, C., and Chang, E. 2007. *Automated Data Capture Technologies—RFID, E-Supply Chain Technologies and Management*. Hershey, PA: IDEA Group Reference.

Radio Frequency Identification news and commentary. 2011. http://www.rfidgazette. org/walmart/ (accessed November 20, 2011).

Ranganathan, C., Dhaliwal, J.S., and Teo, S.H.T. 2004. Assimilation and diffusion of Web technologies in supply-chain management: An examination of key drivers and performance impacts. *International Journal of Electronic Commerce*, 9(1), 127–161.

Scheer, A.W., and Nüttgens, M. 2000. ARIS Architecture and reference models for business process management. *Business Process Management, Lecture Notes in Computer Science*, 1806/2000, 301–304.

Singh, V., and Jenamani, M. 2007. Composite Web services for implementing vendor managed inventory. In *Fifth IEEE International Conference on Industrial Informatics*, Vienna, Austria. June 23–27, 791–795.

Sinkovics, R.R., Jean, R.B., Roath, A.S., and Tamer, C.S. 2011. Does IT integration really enhance supplier responsiveness in global supply chains? *Management International Review*, 51(2), 193–212.

Themistocleous, M., Irani, Z., and Love, P.E.D. 2004. Evaluating the integration of supply chain information systems: A case study. *European Journal of Operational Research*, 159(2), 393–405.

Vernadat, F.B. 2002. Enterprise modeling and integration (EMI): Current status and research perspectives. *Annual Reviews in Control*, 26(1), 15–25.

Wang, W.Y.C., and Chan, H.K. 2010. Virtual organization for supply chain integration: Two cases in the textile and fashion retailing industry. *International Journal of Production Economics*, 127(2), 333–342.

Wong, C.Y., Boon-itt, S., and Wong, C.W.Y. 2011. The contingency effects of environmental uncertainty on the relationship between supply chain integration and operational performance. *Journal of Operations Management*, 29(6), 604–615.

Yi-nan Qi., and Zhao-fang Chu. 2009. The impact of supply chain strategies on supply chain integration. *Proceedings of International Conference on Management Science and Engineering.* New York: IEEE. DOI: 10.1109/ICMSE.2009.5317307, 534–540.

You, F., and Grossmann, I.E. 2008. Design of responsive supply chains under demand uncertainty. *Computers and Chemical Engineering,* 32, 3090–3111.

Zeng, A.Z., and Pathak, B.K. 2003. Achieving information integration in supply chain management through B2B e-hubs: Concepts and analyses. *Industrial Management and Data Systems*, 103(9), 657–665.

Index